DATA = EQUATIONS

How To Write Equations That Fit The Data

By

Robert Freund

e-mail:robertefreund@sbcglobal.net

ISBN 10 149288345X
ISBN 13 9781492883456

TABLE OF CONTENTS

PREFACE

I have written this book to close the gap between the theoretical treatment of data and the practical application of the data. Specifically, after an experiment has been performed and data is collected this work addresses the practical methods of writing equations that will describe the experimental results.

I received my education in physics and economics at the Universities of Pittsburgh and Miami and received my Ph. D. from Duke University. It was only during the Ph.D. program's work on original research was I confronted with the problem of the mathematical treatment of experimental data.

Only then did I become aware of the many equations that could be written to describe one set of data. Over the years I found more practical ways to solve problems. I became very interested in least squares techniques and how easy the solution became using matrix algebra. A multi-variable relationship could be quickly solved by a simple system of simultaneous equations.

This book is not written for mathematicians. There are no proofs, discussions of stability in the section on differential equations. There is no worry about rounding errors. Although these topics are all important, they mean nothing to the beginning engineer or research worker until he or she can first derive a working equation to describe a set of data.

My goal is to provide a set of guidelines to various mathematical techniques and a general methodology to enable the researcher
to mathematically describe his or her experimental results.

Chapter 1. Introduction

The analysis of experimental data and the subsequent derivation of a mathematical equation relating the data's variables is one of the most common and important applications of mathematics. The usual treatment of this subject begins with a study of various interpolation formulas, a study of least squares techniques, and concludes with special problems of simultaneous equation systems. The derivation of a mathematical function based upon the data is usually treated as a minor or obvious problem. The derivation of the mathematical function is the primary focus of this book. The approach used will be one of practical application working from the data to the subsequent mathematics. Such an approach will clearly demonstrate the limits of the various techniques.

BASIC PROBLEM OF DERIVING A FUNCTION

The reader may think the basic task is the construction of a mathematical equation that explains the experimental data satisfactorily. In fact, many equations can be constructed to explain the same data. The most basic problem of analysis, therefore, is to identify the most applicable form the mathematical expression should take.

The form of the mathematical expression refers to how the variables are related to each other. For example, in

$$(1.1) \quad y = a_0 + a_1 x_1 + a_2 x_2$$

$$(1.2) \quad y = a_0 e^x$$

$$(1.3) \quad y = (a_1 x_1)(a_2 x_2)$$

Equation (1.1) assumes the variables x_1 and x_2 are independent from each other's influence on y. Furthermore, these expressions assume that y can be uniquely explained by x_1 and x_2 and the parameters a_0, a_1, and a_2 are constants. In practice this is usually not true. It may be that one of the variables is a satisfactory proxy for a third unknown variable.

The underlying theory of the field investigated (economics, physics, electronics, etc.) guides the investigator in selecting the variables for study or experiment but one is never certain he has explained the true relationship.

The constancy of the parameters a_i is critical when the linear form of (1.1) is used. Subsequent experiments with new data may result in new values of a_i. Are these changes in the value of the parameters due to new influences from other unobserved variables or are they more proper estimates than the previous experimental results? These are questions the analyst must ask.

OBJECTIVES OF THE ANALYSIS

To answer properly the question that arises in function estimation requires a clear understanding of the objectives of the analysis. An engineer calculating the fuel tank size required of a new boat hull design will be satisfied with empirical results over the acceptable r.p.m. range of the engines used.
A sales manager attempting to estimate future sales may use a statistically derived formula only as a guide to check against his experience and intuitive estimates.   Whereas the physicist is more interested in learning the relationship of the variables in a fluid dynamic experiment, the weather forecaster is concerned with a 24 to 48 hour weather forecast with some degree of

accuracy. The economist is concerned with the general law answering why and how much consumers purchase but the sales manager may be fired if his forecasts are inaccurate. The point is that the objective determines to a great extent the form of the solution; and within their proper scope many methods may be correct.

GRAPHS AND DATA CONVERSION

The three most common forms that graphed experimental data take are linear, power or exponential, and sinusoidal functions. Therefore, when one deals with two variables, plotting the data in graphical form is usually the first step. But there remains to be answered a number of questions concerning the proper form the relationship shall take.

Assume the following data:

TABLE 1.1

X	Y
0.00	1.00
0.69	2.00
1.10	3.00
1.39	4.00
1.61	5.00
1.79	6.00
1.95	7.00
2.08	8.00
2.20	9.00
2.30	10.00

When graphed, this data produces an exponential function relating X and Y. The data is from a natural logarithms table and

$$(1.4) \quad Y = e^X$$

$$\text{or } (1.5) \quad \ln Y = X$$

The relationship of (1.5) is frequently used to convert data to the logarithmic form to simplify the analysis. Since the values of Table 1.1 are taken from a table of natural logarithms the conversion results in a one-to-one relationship.

The conversion of data to the logarithmic form is particularly useful when several independent variables jointly influence Y.

For example,

(1.6) $Y = (X_1)^a(X_2)^b$

(1.7) $\ln Y = a(\ln X_1) + b(\ln X_2)$

The form expressed by (1.7) allows the use of least squares techniques to solve for parameters a and b. These techniques will be explained in a later chapter.

If we modify three entries in the data of TABLE 1.1 as follows,

TABLE 1.2

X	Y
0.60	1
0.69	2
1.10	3
1.20	4
1.61	5
1.79	6
1.80	7
2.08	8
2.20	9
2.30	10

We can no longer "fit" the data uniquely with an exponential function. The data can now be

explained by either an exponential or a linear function.
See Figure 1.1. Which is correct?

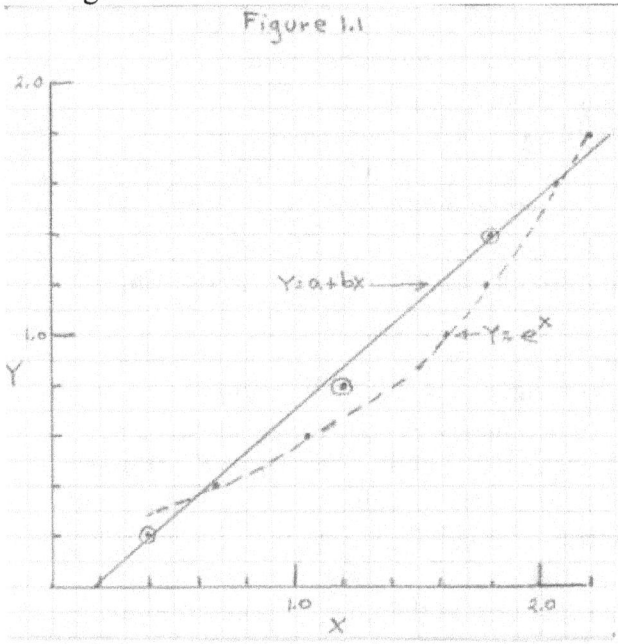

Figure 1.1

THE ART OF CURVE FITTING

The answer to the question, "Which function is correct?" is: it depends upon your objective. They both may be correct.

Using the data graphed in Figure 1.1. the question to be asked is, "What do the variables represent and what is the purpose of the investigation or experiment?". If the data represents the population growth of a culture over time, it is reasonable to assume an exponential growth law is expected and try $Y = e^x$. The changed observations of Table 1.2 might be due to experimental error and the experiment should be repeated. But if Figure 1.1 depicts units produced versus factory costs an industrial engineer or production manager would most probably accept the linear form, $Y = a + bX$ as a good estimate fully realizing the form may change once the limits of the data are exceeded.

The previous example is to illustrate the limits in applying the mathematics. A mathematician given a set of experimental data but no information describing the nature of the experiment can either derive a number of formulas relating the variables and let the customer make his choice or ignore it as not worth his time. Because the form of the function chosen to relate the data is influenced by the objective and scope of the experiment, curve fitting is sometimes referred to as an art. This is because there is usually no unique relationship to be found. But there are logical procedures and mathematical tests that will guide the analyst in deriving a reasonable function. The balance of this book discusses these tests and procedures beginning with the construction of a table of differences.

Chapter 2. The Use Of Difference Tables

The first step in approximating a function is to construct a difference table. To best understand why this is so we will begin with a known function,

$$(2.1) \quad Y = X^2$$

And substituting in values of X we obtain value of Y as shown in Table 2.1.

TABLE 2.1

X	Y	Y difference	Y (difference)$_2$
0	0		
1	1	1	
2	4	3	2
3	9	5	2
4	16	7	2
5	25	9	2

Using the calculus on (2.1),

$$\frac{dY}{dX} = 2X$$

$$dY = 2XdX$$

$$\int dY = \int 2XdX$$

$$Y = X^2$$

The purpose of the calculus operations is to show that a knowledge of the derivative allows us to reconstruct the original equation. With experimental data, however, the derivative is unknown since the function is unknown. Therefore, we construct a difference table from the data to estimate, if possible, the derivative.

If we conclude, since $\Delta X = 1$, that ΔY and $\Delta_2 Y$ are close approximations of the first and second derivatives, then seeing that $\Delta_2 Y$ is constant we can conclude that the variables represented by the data can be related by a second order polynomial equation. Figure 2.1 shows the relationship between the original function and the first and second differences.

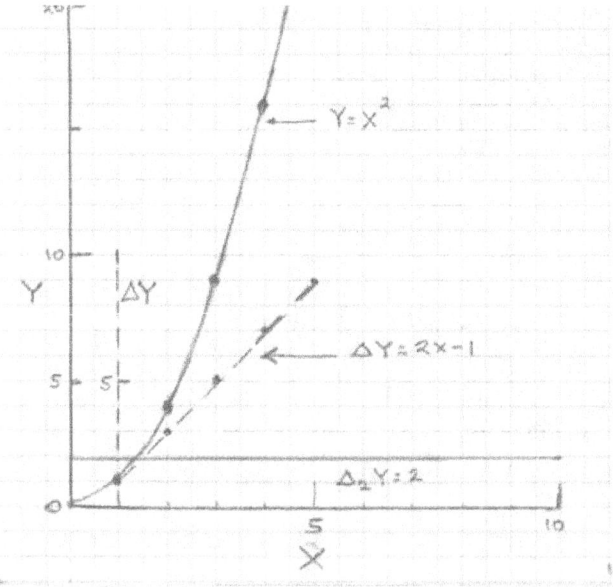

Figure 2.1

Now let us reverse the procedure. Assume an experiment has produced the data of the previous Table 2.1. The appearance of a constant in the $\Delta_2 Y$ column indicates a second order polynomial can relate the variables. Therefore, we assume,

(2.2) $Y = aX^2 + bX + c$

Since we have three unknowns, a,b, and c, we select three data sets of X and Y and then set up three equations and solve for a, b, and c.

Substituting values from Table 2.1,

$$Y \qquad X^2 \qquad X$$

$$(1) = a(1)^2 + b(1) + c$$
$$(4) = a(2)^2 + b(2) + c$$
$$(9) = a(3)^2 + b(3) + c$$

Solving these three equations yields

$$a = 1$$
$$b = 0$$
$$c = 0$$

and substituting into (2.2) gives the original equation

$$Y = X^2$$

This is the basic procedure used when a difference table yields a column of constants. The order of the column is the degree of the polynomial and data sets equal to the order of the column are chosen to enable the solution of the polynomial's parameters.

But experimental data rarely yields results of such simplicity. The data of Table 2.1 is more likely to be,

TABLE 2.2

X	Y	ΔY	$\Delta_2 Y$	$\Delta_3 Y$	$\Delta_4 Y$	$\Delta_5 Y$
0	0.3					
1	1.1	0.8				
2	3.6	2.5	1.7			
3	8.8	5.2	2.7	1.0		
4	15.7	6.9	1.7	-1.0	-2.0	
5	25.4	9.7	2.8	1.1	2.1	4.1

This difference table does not have any column of constants such as found in Table 2.1. This is the usual pattern and is caused by noise, experimental error, or equally likely the data may be noise/error free but simply does not fit a simple mathematical function. We now face the problems discussed in the introductory chapter.

Remember, the function to be derived is to be calculated within the context of the objective of the specific job or experiment being performed. Rushing to the nearest computer terminal will produce a lot of printouts but little understanding. Just as bad is to complicate the problem with unnecessary mathematically sophisticated techniques. The more sensible thing to do is to graph the data and see if we can propose a most probable form the function should have.

A graph of the data, as shown in Figure 2.2, indicates either a power function or an exponential function should apply.

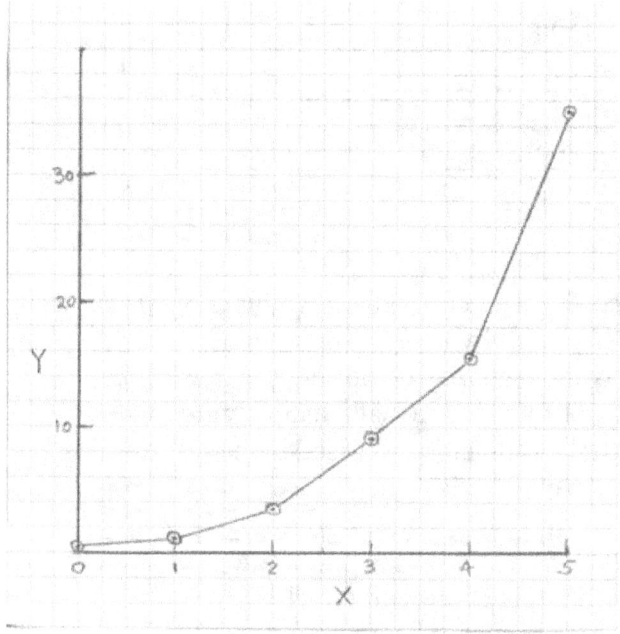

Figure 2.2

To help decide if the data can best be described by a power or exponential function, form Table 2.3.

Table 2.3

X	Y	Y_{n+1}/Y_n
0	0.3	
1	1.1	3.67
2	3.6	3.27
3	8.8	2.44
4	15.7	1.78
5	25.4	1.62

Note the Y_{n+1}/Y_n ratio is continually decreasing. This indicates a power function. If the ratio was constant or, more likely, clustered about a fixed value an exponential function would be appropriate.

Since there are no inflection points in the curve, a second order polynomial may be sufficient. Refer to Table 2.2. Although there are no columns of constants, note that as we pass from Δ_2 to the Δ_3 column the sign changes in some of the differences.
Therefore, a first trial can be to assume a second order polynomial.

$$(2.4\} \quad Y = aX^2 + bX + c$$

and substitute three sets of data.

$$1.1 = a(1)^2 + b(1) + c$$
$$3.6 = a(2)^2 + b(2) + c$$
$$8.8 = a(3)^2 + b(3) + c$$

Solving,

$$a = 1.35$$
$$b = -1.55$$
$$c = 1.30 \text{ and substituting in (2.4)}$$

$$\tilde{Y} = 1.35x^2 - 1.55x + 1.30$$

The symbol $\tilde{Y}$ represents the calculated value and Y the observed value. If we let $e = Y - \tilde{Y}$, we can measure the sum of the differences squared, or error, of $\tilde{Y}$ from Y.

TABLE 2.3

Y	.30	1.10	3.60	8.80	15.70	25.40
$\tilde{Y}$	1.30	1.10	3.60	8.80	16.70	27.30
e	-1.0	0	0	0	-1.0	-1.9

$\Sigma e^2 = 5.61$

Since the emphasis of this book is on the practical applications of elementary numerical analysis, let's stop and review what we have done.

1. Constructed a difference table and from it and the shape of the sketched curve estimated a second order polynomial will be the minimum satisfactory approximation.

2. To demonstrate the basis of the technique we constructed three linear
equations and solved them simultaneously to obtain

$$Y = 1.35x^2 - 1.55x + 1.30$$

With $\Sigma e^2 = 5.61$

Now the practical next step to take next is to see if the
derived equation can be simplified.

Assume a = 1, b = 0, and c = 0 then $Y = X^2$. The
results are in Table 2.4

TABLE 2.4

Y	.30	1.10	3.60	8.80	15.70	25.40
Ŷ	0	1.00	4.00	9.00	16.00	25.00
e	.30	.10	-.40	-.20	-.30	.40

$\Sigma e^2 = 0.55$
The sum of the squared errors is less and $Y = X^2$ is
definitely more satisfactory than
$Y = 1.35 x^2 - 1.55 x + 1.30$.

One more example may be helpful in clarifying some
of the methods we have covered so far. The following
is a set of data that is graphed in Figure 2.3 and the
differences in Table 2.4.

FIGURE 2.3

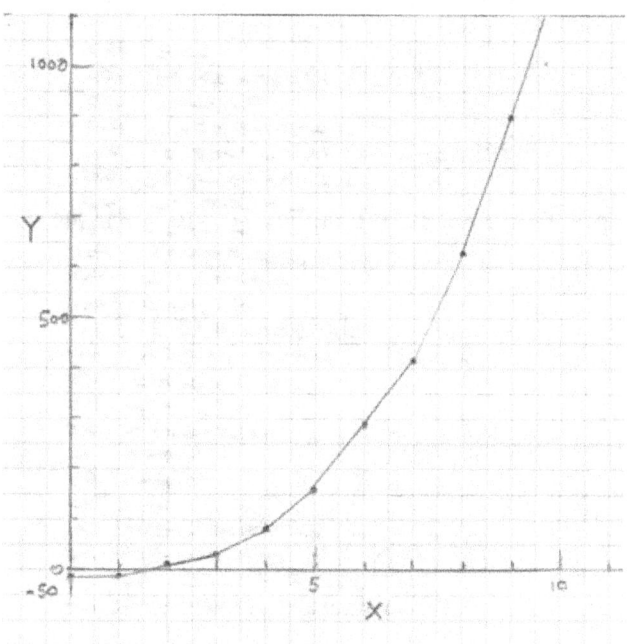

Next a table of differences is constructed.

TABLE 2.4

X	Y	ΔY	$\Delta_2 Y$	$\Delta_3 Y$
0	-15			
1	-10	5		
2	2	12	7	
3	28	26	14	7
4	87	59	23	11
5	160	73	14	-9
6	290	130	57	43
7	415	125	-5	-62
8	630	215	90	95
9	900	270	55	-55
10	1175	275	5	-50

The first step of the analysis indicates a second order polynomial should be tried since a significant sign reversal occurs with third differences. As before we select three data pairs and solve for the parameters of,

$$Y = aX^2 + bX + c$$

$$-15 = a(0)^2 + b(0) + c$$
$$160 = a(5)^2 + b(5) + c$$
$$1175 = a(10)^2 + b(10) + c$$

$$-15 = c$$
$$160 = 25a + 5b + c$$
$$1175 = 100a + 10b + c$$

Solving these three equations gives

$$a = 16.96$$

$$b = -49$$

$$c = -15$$

and

$$\hat{Y} = 16.96X^2 - 49X - 15$$

Using this equation we calculate $\hat{Y}$ and e.

TABLE 2.5

X	Y	$\hat{Y}$	e
0	-15	-15	0
1	-10	-47	37
2	2	-46	44
3	28	-11	39
4	87	58	29
5	160	160	0
6	290	296	-4
7	415	465	-50
8	630	668	-38
9	900	905	-5
10	1175	1175	0

$$\Sigma \, e^2 = 9652$$

The calculated values $\hat{Y}$ do not provide a good fit in the lower half of the range of independent variable X. To improve the fit we can try fitting the next higher order polynomial. To do this we select four variable pairs of data and solve for the parameters of

$$Y = aX^3 + bX^2 + cX + d$$

$$-10 = a(1)^3 + b(1)^2 + c(1) + d$$

$$28 \, = \, a(3)^3 + b(3)^2 + c(3) + d$$

$$290 = a(6)^3 + b(6)^2 + c(6) + d$$

$$900 = a(9)^3 + b(9)^2 + c(9) + d$$

$$-10 = a + b + c + d$$

$$28 = 27a + 9b + 3c + d$$

$$290 = 216a + 36b + 6c + d$$

$$900 = 729a + 81b + 9c + d$$

Solving this 4X4 matrix with a TI calculator capable of matrix operations gives,

$$Y = 0.71X^3 + 6.58X^2 - 16.54X - 0.75$$

TABLE 2.5

X	Y	$\hat{Y}$	e
0	-15	-.75	-14.25
1	-10	-10	0
2	2	-1.8	3.8
3	28	28	0
4	87	83.8	3.2
5	160	169.8	-9.8
6	290	290	0
7	415	449	-34
8	630	652	-22
9	900	901	-1
10	1175	1202	-27

$\Sigma e^2 = 2694$

Since the sum of squared errors is approximately ¼ that of the errors generated by a second power equation, we conclude a third degree polynomial fits the data better.

SUMMARY

1. Graph the data and construct a
. difference table.

2. Estimate the degree of the equation.

3. Select data pairs spaced across the range of the data.

The purpose of this chapter is to introduce to the reader the use of a difference table as well as its limitations. The need for educated interpretation becomes apparent when more realistic data sets are analyzed. The more realistic data reduces the role of the difference table to that of a guide in determining the most appropriate degree of the approximating polynomial.

These illustrations are also intended to show that much applied analysis can be done without the application of advanced mathematics. The selection of three or four sets of data observation and the construction and solution of a set of linear simultaneous equations to derive an approximating function can produced satisfactory results. Very few realistic applications require fourth, fifth, or higher degree equations to describe the relationship between two variables.

Chapter 2. Problems

1. Given:

X	0	1	2	3	4	5
Y	-3	-3	0	6	12	20

Find a second degree equation to describe the data.

2. Given:

X	Y
1	23
2	10
3	28
4	10
5	20
6	32
7	30
8	58
9	65
10	80

Graph the data. Will a quadratic work? If not, try a cubic.

3. Given:

X	Y
1	9
2	32
3	38
4	34
5	22
6	20
7	28
8	37
9	52
10	75

Graph the data and find an equation.

Solutions.

Problem 1.

Graph data.

Try a second degree equation,

Set up three equations

$$-3 = a_0 + a_1(1) + a_2(1)^2$$
$$6 = a_0 + a_1(3) + a_2(3)^2$$
$$20 = a_0 + a_1(5) + a_2(5)^2$$

$$-3 = a_0 + a_1 + a_2$$
$$6 = a_0 + 3a_1 + 9a_2$$
$$20 = a_0 + 5a_1 + 25a_2$$

Solving,

$$a_0 = -5.625$$
$$a_1 = 2.000$$
$$a_2 = 0.625$$

$$\hat{Y} = -5.625 + 2X + 0.624\ X^2$$

X	0	1	2	3	4	5
Y	-3	-3	0	6	12	20
$\hat{Y}$	-5.625	-3	.875	6	12.375	20

Problem 2.

Graphed data indicates a quadratic might not work. Try a cubic.

Set up 4 equations.

$$10 = a_0 + a_1(2) + a_2(2)^2 + a_3(2)^3$$
$$20 = a_0 + a_1(5) + a_2(5)^2 + a_3(5)^3$$
$$58 = a_0 + a_1(8) + a_2(8)^2 + a_3(8)^3$$
$$80 = a_0 + a_1(10) + a_2(10)^2 + a_3(10)^3$$

$$10 = a_0 + 2a_1 + 4a_2 + 8a_3$$
$$20 = a_0 + 5a_1 + 25a_2 + 125a_3$$
$$58 = a_0 + 8a_1 + 64a_2 + 512a_3$$
$$80 = a_0 + 10a_1 + 100a_2 + 1000a_3$$

Solving,
$$a_0 = 37.8$$
$$a_1 = -23.1$$
$$a_2 = 5.1$$
$$a_3 = -0.24$$

$$\hat{Y} = 37.8 - 23.1X + 5.10X^2 - 0.24X^3$$

X	1	2	3	4	5	6	7	8	9	10
Y	23	10	28	10	20	32	30	58	65	80
$\hat{Y}$	20	10	8	12	20	31	44	57	60	77

Problem 3.

Given:

X	1	2	3	4	5	6	7	8	9	10
Y	9	32	38	34	22	20	28	37	52	75

The graphed data shows two inflection points at X= 3 and X=6.

This would normally raise the question of more than one independent variable acting upon the dependent variable Y. But for this problem let's remain with one independent variable and <u>select four data pairs upon each side of the inflection points.</u>

$$9 = a_0 + a_1(1) + a_2(1)^2 + a_3(1)^3$$
$$38 = a_0 + a_1(3) + a_2(3)^2 + a_3(3)^3$$
$$20 = a_0 + a_1(6) + a_2(6)^2 + a_3(6)^3$$
$$52 = a_0 + a_1(9) + a_2(9)^2 + a_3(9)^3$$

$$9 = a_0 + a_1 + a_2 + a_3$$
$$38 = a_0 + 3a_1 + 9a_2 + 27a_3$$
$$20 = a_0 + 6a_1 + 36a_2 + 216a_3$$
$$52 = a_0 + 9a_1 + 81a_2 + 729a_3$$

$$a_0 = -33.30$$
$$a_1 = 54.10$$
$$a_2 = -12.70$$
$$a_3 = 0.86$$

$$\hat{Y} = -33.30 + 54.10X - 12.70X^2 + .86X^3$$

X	1	2	3	4	5	6	7	8	9	10
Y	9	32	38	34	22	20	28	37	52	75
$\hat{Y}$	9	31	38	35	27	20	18	27	52	98

This provides a good fit but if the experiment warranted it, other independent variables should be considered and multiple least square regression tried (see Chapter 5).

Chapter 3. Taylor Expansions And
Lagrangian Equations

The Taylor expansion theorem can be expressed in a
form referred to as the Gregory-Newton formula,

$$(3.1)\ Y = f(0) + X(\Delta f) + \underline{X(X-1)}(\Delta_2 f)/2!$$
$$+ \underline{X(X-1)(X-2)}(\Delta_3 f)/3!+$$

This formula is most often used for
interpolation between two given numbers within a set
of data. But given a relatively well ordered data set it
can be used with a difference table to immediately
write down the equation to describe the function.
But remember this caution: Do not extrapolate beyond
the data set.

Consider the following,

TABLE 3.1

X	Y	ΔY	$\Delta_2 Y$
0	-25	2	2
1	-23	4	2
2	-19	6	2
3	-13	8	2
4	-5	10	2
5	5	12	2
6	17	14	2
7	31	16	2
8	47		

We can now substitute from this table directly into formula (3.1) and solve for the resulting function. Substituting the first entry of each column across the X = 0 row gives,

$$Y = (-25) + X(2) + \frac{X(X-1)(2)}{2!}$$

$$Y = -25 + 2X + \frac{2X^2 - 2X}{2}$$

$$Y = X^2 + X - 25$$

This quadratic equation has been calculated from the data without the need for solving three simultaneous linear equations. You can also start at other points in the table. For example, starting at row x = 5,

$$Y = 5 + (X-5)12 + \frac{(X-5)((X-5)-1)(2)}{2!}$$
$$Y = 5 + 12X - 60 + X^2 - 11X + 30$$
$$Y = X^2 + X - 25$$

the same result as before.

When compared with our previous examples of
building a difference table, constructing and solving
3X3 or 4X4 matrices, this is a definite simplification.
The clean and accurate results are, of course, due to
our selection of data for this illustration. Nevertheless,
what is shown is a very fast and simple method,
suitable for pocket calculator application, that one can
use as a first step in the approximation of a function.

Let us summarize what we have done so far

> 1. We have learned how a difference table
> may be used not only as a guide in
> the degree best suited for a polynomial
> expansion, but it can provide directly
> the inputs needed for a first approxi-
> mation of the function.
> 2. The Gregory-Newton formula can be
> used for a fast estimate of the suitabil-
> ity of the equation considered as the
> estimating function.
> 3. We need only select the mid-point data
> and with a minimum number of cal-

culations have a very quick polynomial estimating function.
4. Adjust the differences and test their effects.

Assume the four steps above have been done and a quadratic function gives generally reasonable results. The next step is to collect a new data set from a repeated experiment and confirm a quadratic function is still suitable. If so, what are the the theoretical ties between the variables that may explain a second degree relationship? If all these steps lead to positive answers and reasonable explanations, then we can consider if it is necessary to perform a least-squares regression.

The reason I believe these steps should precede least-squares is the abuse of this procedure due to the increased availability of computer terminals. The preparation of data input and program formatting take more time than the above four steps. Furthermore, no attempt is usually made to understand the problem. Then when the researcher discovers linear model building and two-stage least squares, chaos reigns. Keep things simple.

Now we must consider how to treat data if the values observed of the independent variable, X, are not equally spaced. There are two common solutions. The first is the method described in Chapter 2: select the required data sets and solve the resultant simultaneous equations. The second is the Lagrangian technique. To illustrate both methods we will use the data from the previous example but omit one observation.

TABLE 3.2

X	5	6	8
Y	5	17	47

The X values are no longer evenly spaced and a normal difference table cannot be constructed. The solution of three linear equations, assuming we have concluded a quadratic equation will be used to approximate the function, is required.

$$a(5)^2 + b(5) + c = 5$$

$$a(6)^2 + b(6) + c = 17$$

$$a(8)^2 + b(8) + c = 47$$

expanding,

$$25a + 5b + c = 5$$

$$36a + 6b + c = 17$$

$$64a + 8b + c = 47$$

and solving,

$$a = 1$$
$$b = 1$$
$$c = -25$$

which gives,

$$Y = X^2 + X - 25$$

the same result found by the Gregory-Newton formula. Now we repeat the same example using the Lagrangian method.

The general quadratic expression is,

(3.3) $Y = aX^2 + bX + c$

and we are attempting to pass this quadratic through three given points (Y_0, X_0), (Y_1, X_1), and (Y_2, X_2). Therefore, we can express (3.3) in terms of the three pairs of selected data.

$$Y_0 = a(X_0)^2 + bX_0 + c$$

$$Y_1 = a(X_1)^2 + bX_1 + c$$

$$Y_2 = a(X_2)^2 + bX_2 + c$$

Next, we define

$$P_0(X) = (X - X_1)(X - X_2)$$

$$P_1(X) = (X - X_0)(X - X_2)$$

$$P_2(X) = (X - X_0)(X - X_1)$$

then (3.3) can be expressed as a weighted sum of the above polynomials

(3.4) $Y = b_0 P_0(X) + b_1 P_1(X) + b_2 P_2(X)$

where,

$$b_0 = Y_0/P_0(X_0)$$

$$b_1 = Y_1/P_1(X_1)$$

$$b_2 = Y_2/P_2(X_2)$$

and substituting into (3.4) gives,

$$Y = Y_0\frac{(X-X_1)(X-X_2)}{(X_0-X_1)(X_0-X_2)} + Y_1\frac{(X-X_0)(X-X_2)}{(X_1-X_0)(X_1-X_2)} +$$

$$Y_2\frac{(X-X_0)(X-X_1)}{(X_2-X_0)(X_2-X_1)}$$

Substituting the values from the previous problem,

$$Y = \frac{5(X-6)(X-8)}{(5-6)(5-8)} + \frac{17(X-5)(X-8)}{(6-5)(6-8)} + \frac{47(X-5)(X-6)}{(8-5)(8-6)}$$

expanding and collecting terms gives,

$$= X^2 + X - 25$$

Lagrange found a method of solving a system of equations by using only one equation. The equation must be constrained to pass through a required number of the original data: three for a quadratic, four for a cubic, etc. If the selected data points are widely spaced the function derived will pass through these points but may vary considerably in between. It is also dangerous to extrapolate beyond the end points of the data set.

SUMMARY

1. Graph the data

2. Build a difference table

3. Use the Gregory-Newton formula if the X values are evenly spaced. If not,

4. Solve a system of simultaneous equations; or

5. Use the Lagrangian method

Chapter 3. Problems

1. Given:

X	0	1	2	3	4	5
Y	0	1	5	10	17	23

Use the Gregory-Newton formula.

2. Given:

X	-2	-1	0	1	2	3	4	5	6
Y	0	-1.5	-2	-1.5	0	2.5	4	3.5	2.5

Use the Gregory-Newton formula.

3. Given:

X	-1	0	4	6
Y	-1.5	-2	4	2.5

Apply the Lagrangian method to this unevenly X spaced data.

Chapter 3 Problem Solutions

1.The use of the Gregory-Newton formula is to be able to write down the equation directly from the difference table.

X	Y	ΔY	$\Delta_2 Y$	$\Delta_3 Y$
0	0	1	3	-2
1	1	4	1	1
2	5	5	2	1
3	10	7	1	
4	17	6		
5	23			

Writing directly from row 1,

$$\hat{Y} = 0 + (1)X + \frac{(3)(X)(X-1)}{2!}$$

$$\hat{Y} = X + \frac{3X^2 - 3X}{2}$$

$$\hat{Y} = 1.5X^2 - 0.5X$$

X	0	1	2	3	4	5
Y	0	1	5	10	17	23
$\hat{Y}$	0	1	5	12	22	35

2. Make the difference table.

X	Y	ΔY	$\Delta_2 Y$	$\Delta_3 Y$
-2	0	-1.5	1.0	0
-1	-1.5	-0.5	1.0	0
0	-2.0	0.5	1.0	0
1	-1.5	1.5	1.0	-2.0
2	0	2.5	-1.0	-1.0
3	2.5	1.5	-2.0	1.5
4	4.0	-0.5	-0.5	
5	3.5	-1.0		
6	2.5			

Writing directly from row 5,

$$\hat{Y} = 0 + 2.5(X-2) + \frac{(X-2)((X-2)-1)(-1)}{2!}$$

$$+ \frac{(x-2)((X-2)-1)((X-2)-2)(-1)}{3!}$$

$$= -4 + .67X + X^2 - .17X^3$$

This equation does not provide a good fit but it does have the necessary two inflection points at X=0 and X=4. The Lagrangian shown next does a much better job.

3. The Lagrangian applied to the previous
   problem but with unevenly spaced X values.

X	-1	0	4	6
Y	-1.5	-2	4	2.5

Writing directly into the Lagrangian form,

$$\hat{Y} = (1.5)\frac{(X-0)(X-4)(X-6)}{(-1+0)(-1+4)(-1+6)} +$$

$$(-2)\frac{(X-(-1))(X-4)(X-6)}{(0(-1))(0-4)(0-6)} +$$

$$(4)\frac{(X-(-1))(X-0)(X-6)}{(4-(-1))(4-0)(4-6)} +$$

$$(2.5)\frac{(X-(-1))(X-0)(X-4)}{(6-(-1))(6-0)(6-4)}$$

Collecting terms,

$$\hat{Y} = -.113X^3 + .76X^2 + .27X - 2$$

Yields,

X	-2	-1	0	1	2	3	4	5	6
Y	0	-1.5	-2	-1.5	0	2.5	4	3.5	2.5
$\hat{Y}$	1.4	-1.4	-2	-1.1	.7	2.6	4	4.2	2.6

A much better fit than the previous cubic.

Chapter 4. Least Squares Regression

Use of the Gregory-Newton formula and polynomial equations may be quick methods of of approximations but are not suitable for all functions. Least squares provides a method of approximating those functions studied in previous chapters as well as exponential and trigonometric functions. In addition, least squares allows the use of all the data and not just selected sets.

There are many books published on the theory of lest squares so it will not be our purpose to discuss this or to review tests of significance. We will concentrate on easy mechanics of deriving regression equations.

If we define the error as the difference

$$e_i = Y_i - \hat{Y}_i$$

where $\hat{Y}$ is the calculated value found from the regression equation, then minimizing the value of the sum of the squares of each e_i found for each X_i will be our measure of accuracy of our regression equation. Minimizing this sum

$$\Sigma e_i^2 = \Sigma (Y_i - \hat{Y}_i)^2$$

is the object of least square regression.

It is a method to fit a curve such that the sum of the squared error terms is minimized.

When the calculus is performed the following matrices are derived.

TABLE 4.1

	1	2	3			
n	ΣX_i	ΣX_i^2	ΣX_i^3	...	a	ΣY_i
ΣX_i	ΣX_i^2	ΣX_i^3	ΣX_i^4	...	b	$\Sigma X_i Y_i$
ΣX_i^2	ΣX_i^3	ΣX_i^4	ΣX_i^5	...	c	$\Sigma X_i^2 Y_i$
ΣX_i^3	ΣX_i^4	ΣX_i^5	ΣX_i^6	...	d	$\Sigma X_i^3 Y_i$
. .					. .	. .

For example, to solve for a quadratic by least squares we use Table 4.1 as follows,

$$(n)a + (\Sigma X_i)b + (\Sigma X_i^2)c = \Sigma Y_i$$

$$(4.1) \quad (\Sigma X_i)a + (\Sigma X_i^2)b + (\Sigma X_i^3)c = \Sigma X_i Y_i$$

$$(X_i^2)a + (\Sigma X i^3)b + (\Sigma X_i^4)c = \Sigma X_i^2 Y_i$$

We now substitute from the experimental data and solve the three equations simultaneously to find the values of the parameters a, b. and c. The values of these parameters will be such to minimize the deviations, or errors, from the experimental data of the derived function

$$\hat{Y} = aX^2 + bX + c$$

This is the method of least square regression. The evaluation of the regression and the statistical significance of the parameters can be found in any statistical textbook and will not be discussed here.

The reader should notice the similarity between (4.1) and the earlier methods used in Chapters 2 and 3. The method is identical; only the treatment of the data is different. Both methods require the solution of three equations simultaneously. But whereas our first method and the Lagrangian method required the use of only three data pairs, the least square method uses the input of all the data pairs, or observations. This increase in information input results in improved control of output. What is unique about the least square method is the number of data observations used does not determine the degree of the derived equation. Recall the number of observations used previously determined whether we would derive a first, second, third, etc. degree equation. Consequently, the power of the least square method is this quality: all observations can be used.

We will compare the use of the least square method with the results of a previous example.

TABLE 4.2

X	0	1	2	3	4	5	6	7	8
Y	-25	-23	-19	-13	-5	5	17	31	47

We will now tabulate the above data into the sums shown in the yellow and blue sectors of Table 4.1. With this data we can then set up three equations and solve for the parameters of the derived equation. This will be done in Table 4.3.

TABLE 4.3

n	X_i	X_i^2	X_i^3	X_i^4	Y_i	X_iY_i	$X_i^2Y_i$
9	0	0	0	0	-25	0	0
	1	1	1	1	-23	-23	-23
	2	4	8	16	-19	-38	-76
	3	9	27	81	-13	-39	-117
	4	16	64	256	-5	-20	-80
	5	25	125	625	5	25	125
	6	36	216	1296	17	102	612
	7	49	343	2401	31	217	1519
	8	64	512	4096	47	396	3008
	-------	-------	-------	-------	-------	-----	------
	36	204	1296	8772	15	600	4968

Inserting the values from Table 4.3 into equations (4.1) gives

$$9a + 36b + 204c = 15$$

$$(4.2) \quad 36a + 204b + 1296c = 600$$

$$204a + 1296b + 8772c = 4968$$

and solving (4.2)

$$a = 1$$

$$b = 1$$

$$c = -25$$

or

$$\hat{Y} = X^2 + X - 25$$

which is identical, as expected, to the quadratic found previously by the other two methods using the same data as inputs.

For a more realistic test of least squares we apply it to the data in Table 4.4

TABLE 4.4

X	Y
0	-15
1	-10
2	2
3	28
4	87
5	160
6	290
7	415
8	630
9	900
10	1175

Doing the arithmetic gives the following:

$$\Sigma X_i = \quad 55$$
$$\Sigma Y_i = \quad 3{,}662$$
$$\Sigma X_i^2 = \quad 385$$
$$\Sigma X_i^3 = \quad 3{,}025$$
$$\Sigma X_i^4 = \quad 25{,}333$$
$$\Sigma X_i^5 = \quad 220{,}825$$
$$\Sigma X_i^6 = 1{,}978{,}405$$
$$\Sigma X_i Y_i = \quad 30{,}761$$
$$\Sigma X_i^2 Y_i = \quad 267{,}137$$
$$\Sigma X_i^3 Y_i = 2{,}384{,}975$$

With this data we solve for a cubic

$$n(a)+(\Sigma X_i)b+(\Sigma X_i^2)c+(\Sigma X_i^3)d = \Sigma X_i$$
$$(\Sigma X_i)a+(\Sigma X_i^2)b+(\Sigma X_i^3)c+(\Sigma X_i^4)d = \Sigma X_i Y_i$$
$$(4.3)\ (\Sigma X_i^2)a+(\Sigma X_i^3)b+(\Sigma X_i^4)c+(\Sigma X_i^5)d = \Sigma X_i^2 Y_i$$
$$(\Sigma X_i^3)a+(\Sigma X_i^4)b+(\Sigma X_i^5)c+(\Sigma X_i^6)d = \Sigma X_i^3 Y_i$$

Substituting the sums into (4.3) gives

$$11a+55b+385c+302d=3{,}662$$

$$55a+385b+3025c+25333d=30{,}761$$

$$385a+3025b+25333c+220825d=267{,}137$$

$$3025a+25333b+220825c+1978405d=2{,}384{,}975$$

and solving results in,

$$a = 0.91$$

$$b = 3.21$$

$$c = 3.29$$

$$d = -13.03$$

or,

$$\hat{Y} = 0.91X^3+3.21X^2+3.29X-13.03$$

and solving for e_i^2,

TABLE 4.5

X	Y	$\hat{Y}$	e
0	-15	-13.03	-1.97
1	-10	-12.20	2.20
2	2	.51	1.49
3	28	30.56	-2.56
4	87	83.41	3.59
5	160	164.52	-4.52
6	290	279.35	10.65
7	415	433.36	-18.36
8	630	632.01	-2.01
9	900	880.76	19.24
10	1175	1185.07	-10.07

$\Sigma e_i^2 = 976.95$

The sum of squares of the errors as the single measure for determining which equation is the "best" is too simplistic. Least squares produces the minimum sum of squares of the error terms. This is what it is designed to do. The reseacher must decide if the function is cubic, logarithmic, or some other relationship. Finally, this is one set of data and additional repetitions with new data may be cubic but shifted slightly. New data will require new calculations resulting in new values for the parameters.

The first task is to find a mathematical form that satisfactorily relates the variables. Will the relationship hold over repeated experiments? The importance of the accuracy of the calculated values, $\hat{Y}$, must be evaluated. In the final stages of the investigation, after repeated experiments confirm the relationship, then a least square regression is appropriate. But until this stage is reached all of the earlier work can be easily performed with hand held calculators and can quickly provide workable approximation functions.

LOG TRANSFORMATION

If the plot of the logY versus X results in a close straight line fit the form of the function can be,

(4.5) $Y = ae^X$

To apply least squares we transform

$$Log\ Y = log\ a + X\ log\ b$$

Let

$$Z = log\ Y$$

$$A = log\ a$$

$$B = log\ b$$

then

$$Z = A + BX$$

and using Table 4.1 as a guide,

$$\begin{bmatrix} n & \sum Xi \\ \sum Xi & \sum Xi^2 \end{bmatrix} * \begin{bmatrix} A \\ B \end{bmatrix} = \begin{bmatrix} \sum Zi \\ \sum XiZi \end{bmatrix}$$

To illustrate the procedure, the following data is taken from a natural log table.

n	X	Y	Z=lnY	X^2	XZ
6	0	1.00	0	0	0
	1	2.72	1	1	1
	2	7.39	2	4	4
	3	20.09	3	9	9
	4	54.60	4	16	16
	5	148.41	5	25	25
	-------		-------	-------	-------
Totals	15		15	55	55

Substituting in the matrix,

$$\begin{bmatrix} 6 & 15 \\ 15 & 55 \end{bmatrix} * \begin{bmatrix} A \\ B \end{bmatrix} = \begin{bmatrix} 15 \\ 55 \end{bmatrix}$$

$$6A + 15B = 15$$
$$15A + 55B = 55$$

solving,

$$A = 0$$
$$B = 1$$

giving

$$Z = X$$

and transforming back to the original,

$$\log Y = X$$

$$Y = e^X$$

This simplistic illustration was used so the mechanics of the transformation of the data to log form would be clear.

Along the same line, an illustration of how to treat trigonometric functions should be helpful.
First, when does one use degrees and when radians? Degrees can be used directly when plotted in polar coordinates where a radius sweeps a 360 degree region and the length of the radius measures the strength of the dependent variable at various degree points. A field strength measurement from an electromagnetic source is an example. But if a business cycle followed a sinusoidal pattern one wants to plot dollars versus

time, not dollars versus degrees. To do so one converts the time axis to radians.

By definition

$$360° = 2\pi \text{ radians}$$

which means a circle of unit (1) radius has a circumference length of 2Π. Therefore,

$$1° = \frac{2\pi}{360}$$

Let 360 degrees = T, the period or wavelength, and X be the time period, t. Then ,

$2\pi X/T$ is the general conversion formula.

Let

$$Y = a + b \sin WX$$

where,

$$W = 2\pi/T$$

Assume we have recorded the price of a commodity over twenty months,

t	Y
1	3.25
4	4.00
5	3.00
7	2.00
8	2.00
10	2.90
12	4.10
14	4.00
16	2.25
18	2.15
20	3.00

Table 4.6 contains the required calculations.

TABLE 4.6

t	Wt	sin Wt	Y	$(\sin Wt)^2$	Y(sinWt)
1	.63	.59	3.25	.35	1.91
4	2.51	.59	4.00	.34	2.35
5	3.14	.00	3.00	.00	0.00
7	4.40	-.95	2.00	.90	-1.90
8	5.03	-.95	2.00	.90	-1.90
10	6.28	.00	2.90	.00	0.00
12	7.54	.95	4.10	.90	3.90
14	8.80	.59	4.00	.35	2.35
16	10.05	-.59	2.25	.35	1.32
18	11.31	-.95	2.15	.90	-2.04
20	12.57	.00	3.00	.00	0.00

The sums are

$$\sin Wt = -.73$$

$$(\sin Wt)^2 = 5.00$$

$$Y = 32.65$$

$$Y(\sin Wt) = 3.34$$

Solving in matrix form,

$$\begin{bmatrix} n & \sum sinWt \\ \sum sinWt & \sum (sinWt)^2 \end{bmatrix} \begin{bmatrix} A \\ B \end{bmatrix} = \begin{bmatrix} \sum Y \\ \sum Y(sinWt) \end{bmatrix}$$

$$11A - .73B = 32.65$$

$$-.73A + 5.0B = 3.34$$

gives

$$A = 3.04$$
$$B = 1.11$$

resulting in

$$\hat{Y} = 3.04 + 1.11 \, sinWt$$

The results are plotted in Figure 4.1

FIGURE 4.1

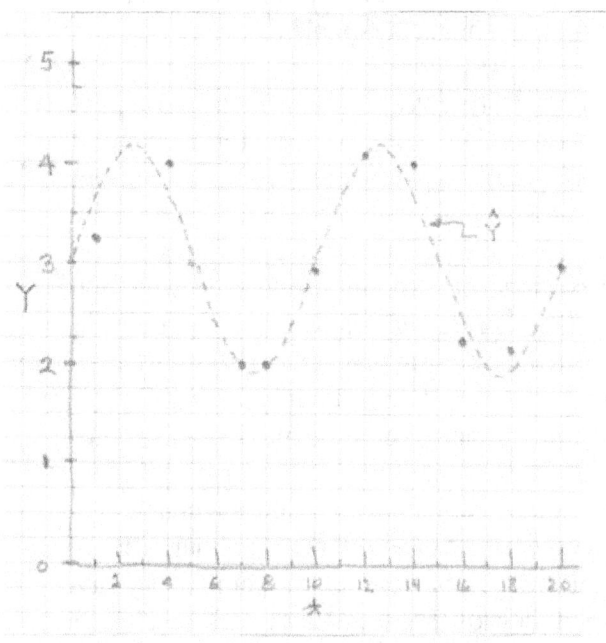

This is a satisfactory results and demonstrates the application of least squares regression to trigonometric functions.

Once again the researcher must be aware that over time the parameters A and B may vary.
It has been stressed since the first chapter that curve fitting is more than a mathematical technique. The approximating function should be explainable by the underlying theory of the appropriate field of study.

It may be helpful to expand this point with a brief reference to the author's personal career.

I have spent thirty five years in the electronics industry, specifically the semi-conductor segment, the fastest growing and most technically innovative sector. I have worked as a production engineer, engineering manager, plant manager, and president. Sales forecasts are regularly made for existing products as well as for new products under development. Price competition is severe and the best selling products are those that were technically impractical two or three years earlier. Now the technical and management people recognize these rapid, almost chaotic, changes but nevertheless must commit millions of dollars for development and production equipment investments. The marketing managers of the various competing companies have all developed their own forecasting techniques varying from elaborate econometric models to outright guesses. The last thing these people need to be told by a newly graduated economist or statistician is the underlying theory is inadequate or the parameters are not statistically significant. They will be told it works today and we will worry about tomorrow the next day. These forecasting techniques, including regression studies, may be sophisticated fortune-telling or tea leaf reading, but given zero data in or being fired going out, some answer must be given. Therefore, in all data interpretation a measure of art or intuition is required. On this note of fortune telling we close this chapter and appropriately turn to a discussion of multiple-regression.

Chapter 4 Problems

1. By least squares derive an equation to describe

X	1	2	3	4	5	6	7
Y	.5	2	2	2.5	4	4	5

2. Given,

X	0	1	2	3	4	5
Y	.5	1.2	3	9.5	17	23

Decide what degree of an equation is suitable and derive the equation by least squares.

Solutions

1. Graphing indicates a single independent
   variable is appropriate.

Y	X	X^2	XY
0.5	1	1	0.5
2	2	4	4
2	3	9	6
2.5	4	16	10
4	5	25	20
4	6	36	24
5	7	49	35
------------	------------	------------	------------
20	28	140	99.5

$$\begin{bmatrix} 7 & 28 \\ 28 & 140 \end{bmatrix} * \begin{bmatrix} a_0 \\ a_1 \end{bmatrix} = \begin{bmatrix} 20 \\ 99.5 \end{bmatrix}$$

$$a_0 = .07$$
$$a_1 = .70$$

$$\hat{Y} = .07 + .70X$$

X	1	2	3	4	5	6	7
Y	.5	2	2	2.5	4	4	5
$\hat{Y}$	.8	1.5	2.2	2.9	3.6	4.3	5

Solutions

2. Graphing indicates a second degree equation
   should be tried.

Y	X	X^2	X^3	X^4	XY	X^2Y
0.5	0	0	0	0	0	0
1.2	1	1	1	1	1.2	1.2
3.0	2	4	8	16	6.0	12.0
9.5	3	9	27	81	28.5	81.5
17.0	4	16	64	256	68.0	272.0
23.0	5	25	125	625	115.0	575.0
-------	-----	-----	-----	-----	-----	-----
54.2	15	55	225	979	218.7	941.7

Solving $\hat{Y} = a_0 + a_1X + a_2X^2$

$$\begin{bmatrix} 6 & 15 & 55 \\ 15 & 55 & 225 \\ 55 & 225 & 979 \end{bmatrix} * \begin{bmatrix} a_0 \\ a_1 \\ a_2 \end{bmatrix} = \begin{bmatrix} 54.2 \\ 218.7 \\ 941.7 \end{bmatrix}$$

$a_0 = -0.28$
$a_1 = 0.89$
$a_2 = 0.77$

$$Y = -.28 + .89X + .77X^2$$

X	0	1	2	3	4	5
Y	.5	1.2	3	9.5	17	23
$\hat{Y}$	-.28	1.4	4.6	9.3	15.6	23.4

This indicates a second degree equation provides a good fit to the data. More importantly it demonstrates the use of least squares to find power functions that best describe experimental results.

CHAPTER 5. MULTIPLE LEAST SQUARE REGRESSION

There are many cases where the researcher suspects the dependent variable, Y, is a function of several independent variables, X_i. These are expressed as

$$Y = f(X_1, X_2, \ldots, X_n)$$

Some examples are the price of a commodity as a function of its supply and the income of consumers, or the concentration level of an impurity in a crystal diffusion process as a function of temperature and lattice defects. The most simplifying assumption to make is that of a linear relationship,

$$Y = a_0 + a_1X_1 + a_2X_2 + \ldots a_nX_n$$

To illustrate least square multiple regression our first example will be that of two independent variables.

$$Y = a_0 + a_1X_1 + a_2X_2$$

And the matrix form is

$$\begin{bmatrix} n & \sum X_1 & \sum X_2 \\ \sum X_1 & \sum X_1^2 & \sum (X_1)X_2 \\ \sum X_2 & \sum (X_1)X_2 & \sum X_2^2 \end{bmatrix} \begin{bmatrix} a_0 \\ a_1 \\ a_2 \end{bmatrix} = \begin{bmatrix} \sum Y \\ \sum (X_1)Y \\ \sum (X_2)Y \end{bmatrix}$$

Let us assume the following data has been collected.

TABLE 5.1

n	Y	X_1	X_2	X_1^2	X_2^2	X_1Y	X_2Y	X_1X_2
10	100	10	3	100	9	1000	300	30
	110	9	5	81	25	990	550	45
	120	8	4	64	16	960	480	32
	122	8	5	64	25	976	610	40
	135	6	7	36	49	810	945	42
	150	4	9	16	81	600	1350	36
	148	3	6	9	36	444	888	18
	175	1	10	1	100	175	1750	10
	180	1	11	1	121	180	1980	11
	200	.5	14	.25	196	100	2800	7
	----	---	---	---	----	------	------	-----
	1440	50.5	74	372,25	658	6235	11653	271

Making the appropriate substitutions into the above matrix gives,

$$10a_0 \quad 50.5a_1 \quad 74a_2 = 1440$$
$$50.5a_0 \quad 372.25a_1 \quad 271a_2 = 6235$$
$$74a_0 \quad 271a_1 \quad 658a_2 = 11653$$

Solving,

$a_0 = 137.45$
$a_1 = -5.05$
$a_2 = 4.33$

which gives,

$$\hat{Y} = 137.45 - 5.05X_1 + 4.33X_2$$

A common method of determining the amount of variation of $\hat{Y}$ from Y is to determine the <u>coefficient of determination</u> defined as,

$$r^2 = \frac{\text{(total variation)-(unassociated var.)}}{\text{(total variation)}}$$

$$r^2 = \Sigma(Y - Y_{ave})^2 - \Sigma(Y - \hat{Y})^2 / \Sigma(Y - Y_{ave})^2$$

To apply this measurement see Table 5.2.

TABLE 5.2

Y	$\hat{Y}$	$Y-Y_{av}$	$(Y-Y_{av})^2$	$Y-\hat{Y}$	$(Y-\hat{Y})^2$
100	100	-44	1936	0	0
110	114	-34	1156	-4	16
120	114	-24	576	6	36
122	119	-22	484	3	9
135	137	-9	81	-2	4
150	156	6	36	-6	36
148	148	4	16	0	0
175	176	31	961	-1	1
180	180	36	1296	0	0
200	196	56	3136	4	16
------	-------	------	------------	-----	------
1440			9678		118

Yav = 144
Substituting

$$r^2 = (9678-118)/9678 = 0.9878$$

There are many statistical tests that can be applied to regression equations that measure the goodness of fit, r^2 being one, the significance of the a_i terms, and one which tests whether the error terms are normally distributed. These tests can all be found in statistic text books. I will discuss only a few elementary and practical points.

After the coefficients a_i have been found and $\hat{Y}$ has been calculated to see if it reproduces the data reasonably well (r^2), the sign of the coefficients should be checked to see if they are what is expected. In this example, X_1 decreases as Y increases and the sign of a_1 is negative as expected. The sign of a_2 is positive signifying that as X_2 increases so does Y. It is quite possible with multiple regressions to have signs of the coefficients to be opposite from that predicted by theory.

There are many facets of multiple regression theory and tests that can be covered only in a text devoted exclusively to the subject of regression theory. What is attempted here is to show the reader the power of multiple regression to estimate complex data and demonstrate the basic mechanics of deriving a multiple regression equation.

Chapter 5. Problems

1. Given two independent variables.

X_1	1	2	3	4	5
X_2	4	2.8	1.8	1	0
Y	5	4.5	4.8	5	5

This is an interesting problem. Graph the data and note the additive effect of the two independent functions.

2. Given

X_1	4	6	6	8	10	14
X_2	15	11	8	3	3	2
Y	100	70	50	40	30	10

Another set of two in dependent variable data.

Solution

1.

Y	X_1	X_2	X_1^2	X_2^2	X_1X_1	X_1Y	X_2Y
5	1	4	1	16	4	5	20
4.5	2	2.8	4	7.8	5.6	9	12.6
4.8	3	1.8	9	3.2	5.4	14.4	8.6
5	4	1	16	1	4	20	5
5	5	0	25	0	0	25	0
----	----	----	----	----	-----	----	-----
24.3	15	9.6	55	28.1	19	73.4	46.2

Solving,

$$a_0 = -3.46$$
$$a_1 = 1.70$$
$$a_2 = 1.68$$

yielding,

$$\hat{Y} = -3.46 + 1.70X_1 + 1.68X_2$$

Y	5	4.5	4.8	5	5
$\hat{Y}$	4.96	4.64	4.66	5.02	5.04

A very good match with the data.

Solution
2.

Y	X_1	X_2	X_1^2	X_2^2	X_1X_2	X_1Y	X_2Y
100	4	15	16	225	60	400	1500
70	6	11	36	121	66	420	770
50	6	8	36	64	48	300	400
40	8	3	64	9	24	320	120
30	10	3	100	9	30	300	90
10	14	2	196	4	28	140	20
----	----	----	----	----	----	-----	-----
300	48	42	448	432	256	1880	2900

In matrix form,

$$
\begin{matrix}
6 & 48 & 42 \\
48 & 448 & 256 \\
42 & 256 & 432
\end{matrix}
\quad
\begin{matrix}
a_0 \\
a_1 \\
a_2
\end{matrix}
\quad
\begin{matrix}
300 \\
1880 \\
2900
\end{matrix}
$$

Solving,

$$a_0 = 47.89$$
$$a_1 = -3.19$$
$$a_2 = 3.95$$

Giving $\hat{Y} = 47.89 - 3.19X_1 + 3.95X_2$

Y	100	70	50	40	30	10
$\hat{Y}$	94	72	60	34	28	11

CHAPTER 6. CONSTRUCTION OF A DIFFERENTIAL EQUATION

Now that we have studied the basic techniques of forming an equation to approximate experimental data, we will look at the method of constructing a differential equation from such data. Differential equations are referred to as the most powerful of applied mathematical tools. The mathematical texts devoted to the subject concentrate on the methods of solution of various types of differential equations. Rarely does a book say, "Now here is the data and this is how we formulate a differential equation to describe the data." This chapter will take a set of experimental data and find an approximating function and show this in itself is insufficient and experiment further until a differential equation is developed.

Our example will derive Newton's Law of Cooling from a set of experimental data. The data along with the difference is given in Table 6.1

TABLE 6.1

t	T	ΔT	$\Delta_2 T$
0.0	37.0	7.0	-4.0
0.5	44.0	3.0	1.0
1.0	47.0	4.0	-0.5
1.5	51.0	3.5	0
2.0	54.5	3.5	-0.5
2.5	58.0	2.0	+0.5
3.0	60.0	2.5	0
3.5	62.5	2.5	-0.5
4.0	65.0	2.0	-1.0
4.5	67.0	1.0	
5.0	68.0		

This experiment consists of bringing a thermometer from an outside temperature of 37degrees Farhenheit into a room of 76 degrees and reading and recording the temperature every 30 seconds as the thermometer's temperature rises to the equilibrium room temperature.

Figure 6.1is a plot of temperature T versus time.

FIGURE 6.1

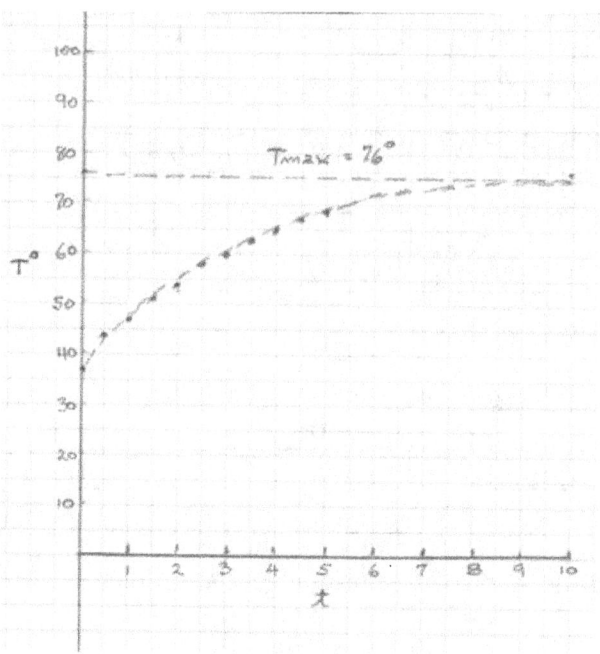

Table 6.1 shows first and second differences are not promising so the natural logs of t and T were tabulated as well as their differences. See Table 6.2

TABLE 6.2

ln t	ln T	Δln T	Δ₂ln T
	3.6109	0.1733	-0.1074
-0.6931	3.7842	0.0659	+0.0158
0.0000	3.8501	0.0817	-0.0153
0.4055	3.9318	0.0664	-0.0042
0.6931	3.9982	0.0622	-0.0293
0.9163	4.0604	0.0339	+0.0070
1.0986	4.0943	0.0409	-0.0017
1.2528	4.1352	0.0392	-0.0089
1.3863	4.1744	0.0303	-0.0155
1.5041	4.2047	0.0148	
1.6094	4.2195		

Figure 6.2 plots lnT versus ln t and the results look promising.

FIGURE 6.2

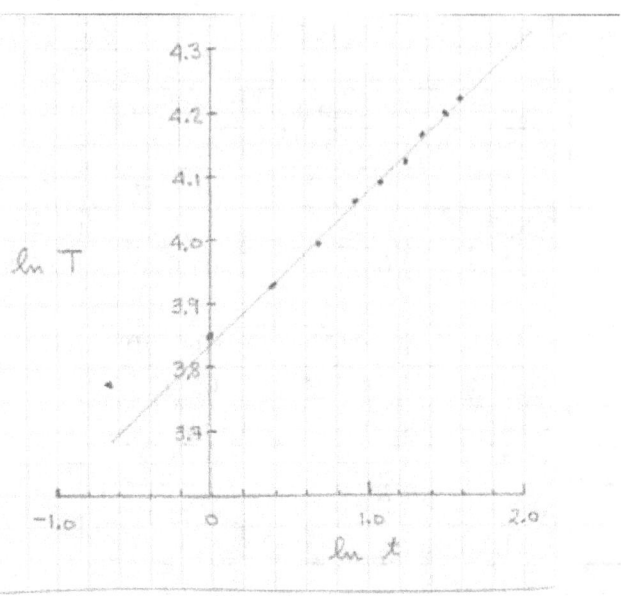

Laying down a ruler and drawing a fairly close fit leads to

$$b = \frac{4.075 - 3.835}{1 - 0} = 0.2400$$

and therefore,

$$\ln T = 3.835 + 0.2400 \ln t$$

If this data were typical of the rise time of a furnace in a production operation, this first approximation is probably as far as one would go. An engineer faced with many problems would be satisfied with this estimated function and it would satisfy the production people working with the furnace. It is an empirical relationship that works for this set of data. But that is also its limitation. It does not explain how the temperature reaches equilibrium. If there were different T maximums, we would need a graph for each T max. This first estimate is not a general solution and the finding of a general solution leads to the construction of a differential equation.

We want an equation that relates the change in temperature with time, dT/dt, as a function of the present temperature and the maximum temperature.

$$\frac{dT}{dt} = f(T, T_{max})$$

How are T and T_{max} related? Figure 6.1 shows that T approaches T_{max} as time increases, reaching T_{max} in approximately 12 minutes. Or, as t increases $(T_{max} - T)$ approaches zero. Let us tabulate these differences at each minute reading of t.

TABLE 6.3

t	$T_{max} - T$
0	39
1	29
2	21.5
3	16
4	11
5	8
...	...
10	1
11	0.5
12	0

and the results are plotted in Figure 6.3.

FIGURE 6.3

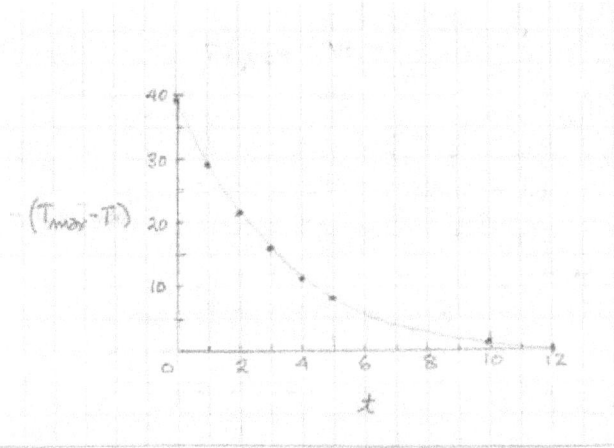

We certainly don't have a simpler relationship than our earlier ln T versus ln t estimate. This leaves us with trying our differential equation

$$\frac{dT}{dt} = \frac{\Delta T}{\Delta t} = f(T_{max} - T)$$

and test to see if some reasonably simple relationship exists between the change in T, ΔT, with t, Δt, versus the decreasing values of T_{max}-T.

We estimate the differential by dividing the values ΔT by Δt as shown in Table 6.1 and calculated below.

TABLE 6.4

t	$\Delta T/\Delta t$	$T_{max} - t$
1	6	29
2	7	21.5
3	4	16
4	5	11
5	2	8
...	...	...
12	0	0

The results are plotted in Figure 6.4

FIGURE 6.4

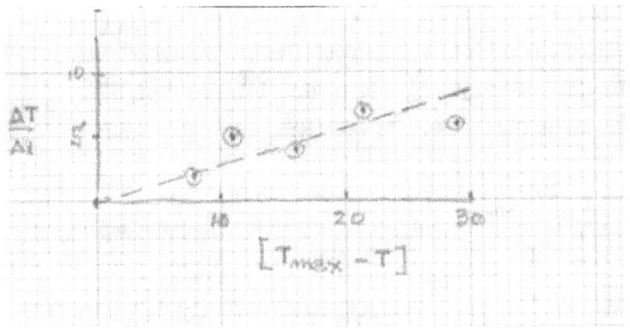

These results are much more encouraging. The fact that the data points do not lie on a perfectly straight line is not surprising. What is significant is we can assume a positive slope, a linear relationship, and the function passes through zero. In other words,

$$Y = bX$$

or in this case,

$$\frac{\Delta T}{\Delta t} = b(T_{max} - T)$$

From Figure 6.4 we can calculate b,

$$b = \frac{7 - 0}{25 - 0} = 0.28$$

Therefore our differential equation can be written as,

$$\frac{dT}{dt} = 0.28(T_{max} - T)$$

This fits our experimental observations in that it says when $T_{max} - T$ is large the rate of change in T is large but when T approaches T_{max} then the rate of change of T approaches zero.

This differential equation is a working equation that reflects correctly the experimental results. It is the second stage of our work. It says the rate of change of the temperature difference decreases as we approach equilibrium but I doesn't say how many minutes are required to reach T_{max}. Nor does it tell us what the temperature will be after so many minutes has passed. To answer these questions requires completing the third stage of the mathematical work, i.e., the development of a general equation expressing T as a function of time.

To do this we use calculus upon our experimentally determined equation.

$$\frac{dT}{dt} = 0.28(T_{max} - T)$$

$$\frac{\int dT}{(T_{max} - T)} = 0.28\int dt$$

$$\ln(T_{max} - T) = 0.28t + c$$

$$T_{max} - T = ce^{-0.28t}$$

$$T = T_{max} - ce^{-0.28t}$$

This is the general solution we have sought and it has been found from experimental data. It states that temperature is a function only of the final temperature T_{max} and time t. We find c by solving the equation when $t = 0$ since $e^0 = 1$.

$$T_0 = T_{max} - c$$

$$c = T_{max} - T_0 = 76 - 37$$

$$c = 39$$

substituting,

$$T = T_{max} - 39e^{-0.28t}$$

The equation is indeed applicable for this experiment. One needs only the initial temperature and the final temperature and the equation gives the temperature at any time. The general expression is,

$$T = T_{max} - ce^{-bt}$$

where,

T_m = maximum or minimum final temperature

$c = T_m - T_0$ (T_0 is T at t = 0)

b = dependent upon rate of change of T to be determined from data

t = time

T = temperature at time t

A summary of the basic steps involved in the construction of a differential equation are:

1. Construct a difference table from the experimental data.

2. Derive a first estimate equation.

3. List the variables of the data as

$$\frac{dY}{dX} = f(X_n)$$

4. Study various relationships of

$$\frac{\Delta Y}{\Delta X} = f(X_i)$$

5. Select a suitable $\frac{\Delta Y}{\Delta X} = f(X_i)$ relationship and solve

$$\frac{dY}{dX} = f(X_i)$$

There are many methods of solving differential equations depending on their form and many cannot be solved analytically but only by numerical approximation. This is a complete field of study in itself and no attempt will be made in this volume of the series to cover it. What has been shown is a straight forward approach to the problem and to show that the first approximation function, although it may fit the experimental data, can be far removed from the final general equation derived from the appropriate differential equation. The decision whether the first approximation is sufficient or whether a more general solution is required depends upon the purpose and scope of the experiment.

Chapter 6. Problems

1. This is a problem in population growth. The data is from the U.S. population growth.

t	P
1	9.6
2	12.9
3	17.1
4	23.2
5	31.4
6	38.6
7	50.2
8	62.9
9	76.0
10	92.0
11	106.5
12	123.2

Graph and then convert P into ln P and graph ln P versus time t. You will see an interesting change in slope between $t = 6$ and $t = 7$.

Solution:

There are two ln P vs t functions.

For t=1 to t=6, $\ln P = 1.97 + .29t$

$$P = 7.17e^{.29t}$$

For t=7 to t=12 $\ln P = 2.7 + .18t$

$$P = 14.88e^{.18t}$$

These two separate equations do an adequate job fitting the data over their respective ranges.

Combining them into a single equation results in what is called the <u>logistic</u> model or equation. It's derivation will be shown in Data = Differential Equations of this series.

SECOND SECTION

DIFFERENTIAL

EQUATIONS

PREFACE

There are no practice problems in this section.
As you read the text you will learn how to create and
solve differential equations from the experimental
data. Practice problems would be artificial and
unnecessary.

All examples in this section were solved using the TI-
83,TI-89, or the TI-200 calculators. .

Chapter 7. Introduction To Differential Equations

The classical approach to differential equations (DE) is to solve a given DE using several methods such as separation of variables, exact solutions using integrating factors, etc. In most real applications the DE cannot be solved directly and approximating methods are used. Known classical solutions can also be compared to the experimental data to see if they give a reasonable approximation of the data.

But given a real application with a set of experimental data, and not given a DE, how does one begin to construct from the data a differential equation that describes the experimental results?

The methodology described here is:

    1. Graph the data Y versus X

    2. Determine the Y_{n+1}/Y_n ratio

        a) If clustered about one value
an
            exponential function is
appropriate

        b) If the ratio is varying a power or
           linear function can be tried

The usual textbook approach starts with a DE and instructs how it can be solved. But for the working scientist or engineer there is no given DE. How does one create a DE to evaluate the experimental data? Follow the methodology discussed above.

The following examples demonstrate how this approach can develop a DE and the function relating the dependent variable, Y, to the independent variable, X.

First, start with a very simple example. Given the data,

$$\begin{array}{ccccccc} X & 0 & 1 & 2 & 3 & 4 & 5 \\ Y & 1 & 2 & 5 & 10 & 17 & 26 \end{array}$$

calculating,

Y_{n+1}/Y_n 2   2.5   2   1.7   1.5

indicates a power function,

and,

$\Delta Y/\Delta X$ 1   3   5   7   9

Plot $\Delta Y/\Delta X$ versus X

See Figure 7.1

Figure 7.1

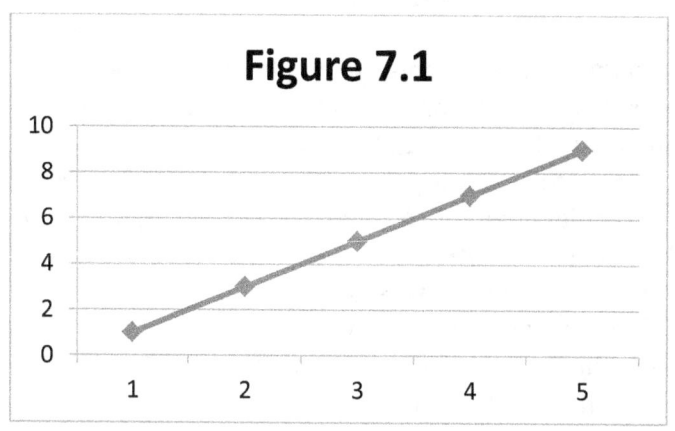

Figure 7.1

then $\Delta Y/\Delta X = (9-1)/(5-1)X + C = 2X + C$

let $\Delta Y = dY$ and $\Delta X = dX$

(1) $dY = 2XdX + CdX$

integrating,

$$Y = X^2 + k$$

at $X = 0$, $Y = 1$ and therefore $k = 1$

and finally,

(2) $Y = X^2 + 1$

A quick check shows (2) provides an exact fit to the data and our DE (1) is correct.

Although this is a very simple example, the method can be applied to many problems that are considerably more difficult. The methodology is:
1. Graph the data
2. Calculate the Y_{n+1}/Y_n ratio
3. Calculate $\Delta Y/\Delta X$
4. Regress $\Delta Y/\Delta X$ versus X
5. Integrate and solve for Y

Today graphic calculators simplify this process and provide great insight into the solution.　　　All the examples in this section were done with the Texas Instrument TI-83+. Texas Instrument Corp. has done an excellent job in support and educational material for it's line of calculators.

The data of the previous example can be made more noisy and, therefore, more realistic. More likely to be obtained from an actual experiment is the following data set,

X	0	1	2	3	4	5
Y	.2	1.8	6	9.5	17	26
$\Delta Y_{n+1}/Y_n$		9	3.3	1.6	1.8	1.5
$\Delta Y/\Delta X$			1.6	4.2	3.5	7.5	9

Figure 7.2

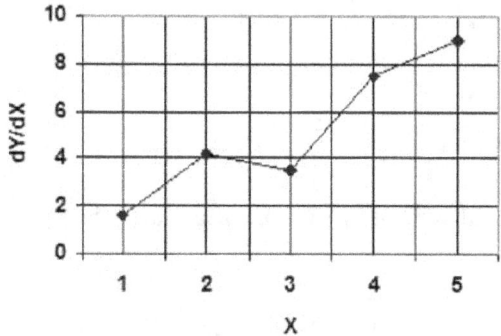

$$\Delta Y/\Delta X = (9 - 1.6)/(5 - 1)X + C = 1.85X + C$$

integrating,

$$Y = 1.85X^2/2 + C = .925X^2 + C$$

$$@X = 1, Y = 1.8, \quad C = 1.8 - .925 = .875$$

$$(2) \quad \hat{Y} = .925X^2 + .875$$

Comparing the actual data with values derived from (3),

X	0	1	2	3	4	5
Y	.2	1.8	6	9.5	17	26
$\hat{Y}$	.88	1.8	4.58	9.2	15.7	24

It is not unreasonable to speculate that after several repeated experiments one might be justified in simplifying,

(3) $\hat{Y} = .925X^2 + .875$
to (4) $\hat{Y} = X^2 + 1$

Let us next deal with a more complex data set.

X	1	2	3	4	5	6	7	8	9	10
Y	.5	.67	.75	.8	.83	.86	.88	.89	.9	.91

$\Delta Y/\Delta X$ .17 .08 .05 .03 .03 .02 .02 .01 .01

Graph the data,

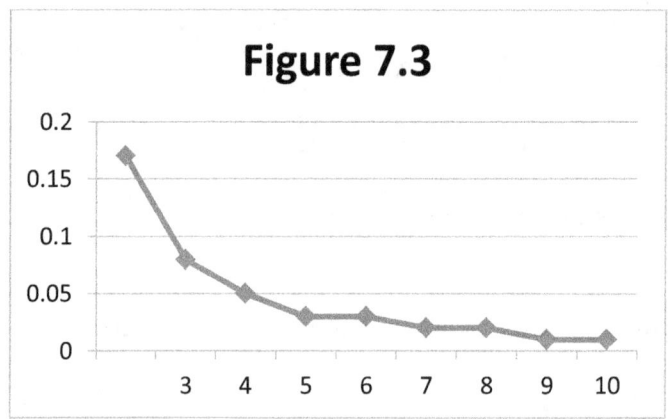

Using a TI calculator a power regression was run yielding,

(4) $\Delta Y/\Delta X = .59X^{-1.8}$
$$Y = .59X^{-.8}/-.8 + C$$

@ X=2, Y=.67

$$.67 = -.74/(2)^{.8} + C$$
$$C = 1.095$$

and, $\hat{Y} = -.74X^{-.8} + 1.095$

As in the previous equation we will modify or "clean up" the derived equation. Let C = 1.

$$\hat{Y} = -.74X^{-.8} + 1$$

But when X = 1, $\hat{Y}$ should equal .5.

Therefore,
$$\hat{Y} = -.5X^{-.8} + 1$$

Finally, so the reader of our report thinks we physicists know what we are doing, make the last change,

$$\hat{Y} = -.5X^{-.75} + 1$$

or even better,

$$\hat{Y} = 1 - (1/2X^{3/4})$$

Now compare the actual Data with the derived $\hat{Y}$.

X	Y	$\hat{Y}$
1	.5	.5
2	.67	.70
3	.75	.78
4	.80	.82
5	.83	.85
6	.86	.87
7	.88	.88
8	.89	.89
9	.90	.90
10	.91	.91

For the record, the original data set was generated by the equation,

$$Y = X/(X+1).$$

Back to a discussion about differential equations and what these examples show us.

First, a working scientist or engineer is never presented with a differential equation. Instead, he or she collects or is given a set of raw data and attempts to understand it. Most likely a function relating the dependent variable to the independent variable is developed and during this stage a DE is derived. The DE solved in the classroom is not found but is good practice.

What is significant in these examples is:

(1) A DE was easily derived from the given experimental data.
(2) This DE was solved for the final Y=f(X).

We have turned upside down the normal method of discussing the solution and derivation of differential equations. The DE was derived from the data and then solved to give an equation that described the given data.

Chapter 8. Velocity Studies

Let us begin by looking at a data set based on Galileo's free fall studies.

t	Y	$\Delta Y_{n+1}/Y_n$	$\Delta Y/\Delta t$
0	0		
1	4.9		4.9
2	19.6	4	14.7
3	44.1	2.25	24.5
4	78.4	1.78	34.3
5	122.5	1.56	44.1
6	176.4	1.44	53.9
7	240.1	1.36	63.7
8	313.6	1.31	73.5
9	396.9	1.26	83.3
10	490.0	1.23	93.1

The Y_{n+1}/Y_n ratio indicates there is no exponential growth.

A TI-83+ linear regression of $\Delta Y/\Delta X$ versus X gives,

$$\Delta Y/\Delta X = -4.9 + 9.8t$$

$$Y = -4.9t + 9.8t^2/2 + C$$

From the data at t = 0 and Y = 0, then C = 0

Also, at t = 1 and Y= 4.9. For this to be true,
-4.9t = 0.
Therefore,

$$Y = 9.8t^2/2$$

Galileo's free fall without friction has been derived from a differential equation developed from the data without any reference to gravity. Of course 9.8 is the gravitational constant and the formula is usually written as,

$$Y = (1/2)gt^2$$

Next the effect of air resistance, or friction, is added. The experimental data is,

t	Y	$\Delta Y/\Delta X$= V	V_{t+1}/V_t
0	0		
1	3	3	
2	16	13	4.33
3	37	19	1.46
4	67	30	1.58
5	104	37	1.156
6	146	42	1.135
7	192	46	1.095
8	241	49	1.065
9	293	52	1.061
10	346	53	1.019
11	401	55	1.037

12	456	55	1.000
13	512	56	1.018
14	568	56	1.000
15	625	57	1.018
16	682	57	1.000
17	736	54	.947
18	795	59	1.093

Graphing,

Figure 8.1

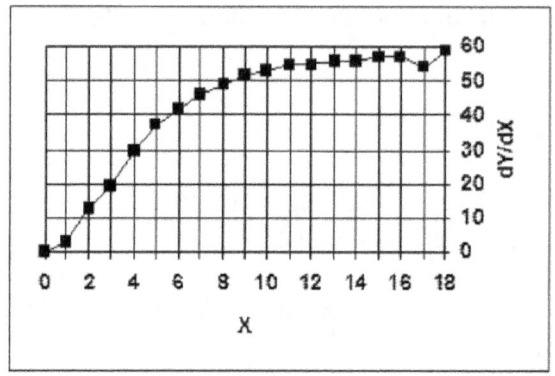

From the graph it is clear that

$$dV/dt = gt - K$$

and an exponential function should describe the data. Note that $V_{max} = 58$. We can , therefore, write

$$dV/dt = k(V_{max} - V_t)$$

Now plot $(V_{max} - V_t)$ versus

t	V	$\Delta Y/\Delta t$	$V_{max} - V$
1	3		55
2	13	10	45
3	19	6	39
4	30	11	28
5	37	7	21
6	42	5	16
7	46	4	12
8	49	3	9
9	52	3	6
10	53	1	5
11	55	2	3
12	55	0	2
13	56	1	2
14	56	0	2
15	57	0	1
16	57	0	1
17	54	-3	4
18	59	5	-1

Graphing the above data,

Figure 8.2

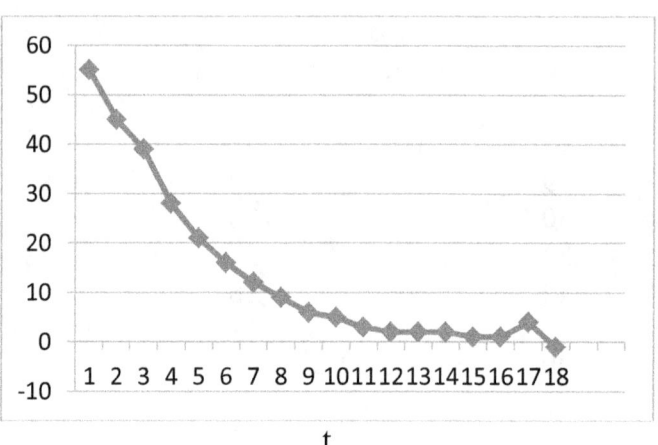

The slope is -.27. It is negative because dV/dt is decreasing.

$$dV/dt = -.27(V_{max} - V)$$

$$V = V_{max} - ke^{-.27t}$$

at $t = 0$, $V = 0$, $k = 58$

$$V = 58 - 58e^{-.27t}$$

$$V = 58(1 - e^{-27t})$$

Chapter 9. The Logistic Equation

In the last chapter we introduced the effect of friction acting upon a free falling body. In population studies a similar effect occurs. The scarcity of resources eventually exerts its influence and the population growth decreases and finally stablizes at a maximun value, P_{max}. Rather than referring to this opposition to growth as friction, the slowing process is usually attributed to "crowding factors". The simple growth model,

$$dP/dX = kP$$

is modified by the percent of resources not used

$$(1 - P/P_{max})$$

to give

(1) $\qquad dP/dX = kP(1 - P/P_{max})$

(2) $\qquad P = (1 - P/P_{max})ke^{X}$

Equation (2) is the logistic equation.

All seems straight forward until one tries to determine P_{max}.

P_{max} is in almost all cases a moving target. It can shift upwards due to technological improvements, for example in food production. It can move down due to wars or natural catastrophes. The ability to fit a curve to the data depends on how accurately P_{max} is determined. What follows is the classical version of the Pearl-Verhulst (as the equation is sometimes called) solution to P_{max}.

X	P	P_{n+1}/P_n	$\Delta P/\Delta X$	$\ln(\Delta P/\Delta X)$
1	.20			
2	.5	2.50	.30	-1.204
3	1.00	2.00	.50	-0.693
4	2.25	2.25	1.25	0.223
5	5.00	2.22	2.75	1.012
6	7.75	1.55	2.25	0.811
7	8.80	1.14	1.05	0.049
8	9.50	1.08	.70	-0.357
9	9.80	1.03	.30	-1.204
10	10.00	1.02	.20	-1.609

Figure 9.1

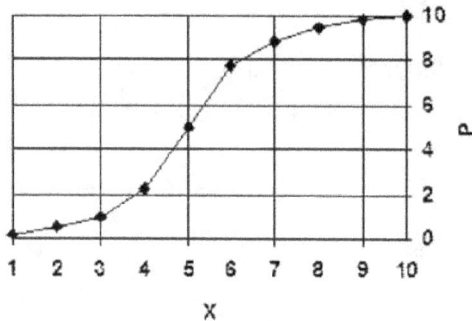

Referring to the above graph, the curve up to the inflection point can be expressed as,

$$dP/dX = P$$

After the inflection point the equation needs to be modified to reflect the decreasing growth rate. This decrease is usually attributed to "crowding factors" represented by,

$$a - bX$$

and combining gives,

$$dP/dX = P(a - bX)$$

Integrating and solving,

$$P = k/1 + e^{(d - at)}$$

where $d = \ln(bk)$, $k = a/b$

Note that as t increases e^{-at} approaches zero and P approaches P_{max}.

Therefore,

$$P_{max} = a/b$$

To solve[1] we need to find a,b,c, and k.

$$k = P_1(2P_0P_2 - P_0P_1 - P_1P_2)/P_0P_2 - P^2$$

$$A = \ln((k - P_0)/P_0)$$

$$B = \ln((k - P_1)/P_1)$$

$$a = (B - A)/(X_1 - X_2)$$

$$d = aX_0 + A$$

1. See Michael Olnick, "An Introduction to Mathematical Models in the Social and Life Sciences", Addison-Wesley, 1978

Using these formulas gives,

   $k = 11.41$

   $A = 4.03$

   $B = 1.40$

   $a = 0.88$

   $d = 4.03$

Substituting into (1),

   $$P = 11.41/(1 + e^{(4.03 - .88X)})$$

Note that as X approaches infinity, P approaches k.
Comparing the actual P with the calculated P_c,

X	P	$\hat{P}_c$
1	.20	.47
2	.50	1.07
3	1.00	2.28
4	2.25	4.28
5	5.00	6.75
6	7.75	8.87
7	8.80	10.20
8	9.50	10.87
9	9.80	11.18
10	10.00	11.31

This is the result of the classical solution. The results are not good and we will discuss why shortly. But first we will look at another approach. This approach uses the methodology developed earlier.

$\ln(\Delta P/\Delta X)$ versus X is listed in Table 9.1

Table 9.1

X	$\ln(\Delta P/\Delta X)$
2	-1.20
3	-0.69
4	0.22
5	1.01
6	1.01
7	0.05
8	0.36
9	-1.20
10	-1.69

We find from the graph there are two functions: one from X 2 to 5 and the other is X 6 to 10.

The calculated results are

$$Y_{1-5} = .11e^{(.76X)}$$
$$Y_{6-10} = 10.5 - 2.7e^{(4.41 - .61X)}$$

My conjecture why the classical method results in not a good fit is that the more curvature on each side of the inflection point of the population curve the worse the accuracy of the data fit.

Chapter 10. Mixed Functions and Methods

Recall in the first chapter the description of mixed functions:

$Z = XY$, X and Y are not independent of each other and are multiplied together.

$Z = X+Y$, X and Y are independent of each other and are added together.

Professor B. West[1] gives an excellent illustration of the additive effect in the book Differential Equation Models. Her approach is to start with the differential equation
$$dY/dT = 2T - Y$$

and graph the field of the slopes of dY/dT. This can easily be done with a TI-89 calculator. My approach is to start with the data relating
$$Y = f(T).$$

The data is,

T	Y	ΔY/ΔT	lnY
- 3.0	52.26		3.96
-2.5	29.55	-45.42	3.39
- 2.0	16.17	-26.76	2.78
- 1.5	8.45	-15.44	2.13
- 1.0	4.15	- 8.50	1.57
- 0.5	1.95	- 4.40	
0.0	1.00	- 1.90	
0.5	0.82	- 0.36	
1.0	1.10	0.56	
1.5	1.67	1.14	
2.0	2.40	1.46	
2.5	3.24	1.68	
3.0	4.14	1.82	
3.5	5.09	1.88	
4.0	6.05	1.92	
5.0	8.00	2.00	
6.0	10.00	2.00	
7.0	12.00	2.00	
8.0	14.00	2.00	
9.0	16.00	2.00	
10.0	18.00	2.00	

The graphed data is shown in the figure below.

Figure 10.1

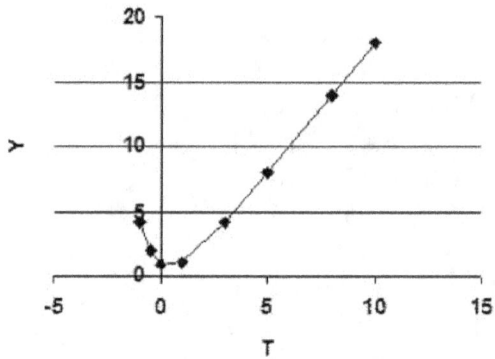

From the graph it is clear there are two functions.

linear from T≥5
non-linear from T≤5

First calculate the linear function,

$$Y = 2T - 2$$

and by inspection we can see the DE is,

$$dY/dT = 2T - Y$$

Next, calculate the non-linear function from T = -3 to T = -1

A linear regression of lnY vs. T gives,

$$\ln Y = -1.2T + 3.5$$

$$Y = e^{-1.2T}e^{c}$$

simplify to,

$$Y = Ae^{-T}$$

Adding the two functions,

$$Y = Ae^{-T} + 2T - 2$$

Solve for A at T=1 and Y = 1.1

$$1.1 = Ae^{-1} + 2 - 2 = Ae^{-1} = .37A$$

$$A = 2.97$$

let A = 3

Finally,

$$Y = 3e^{-T} + 2T - 2$$

Professor West sketched the field slopes of the DE and analyzed the equation qualitatively.
Our approach worked from the data to derive the DE and solve for the dependent variable Y.

Here is another example.

Assume you are given the following data,

T	Y	$\Delta Y/\Delta T$
-2.0	-3.53	
-1.5	-0.83	5.40
-1.0	-0.10	1.46
-0.5	0.29	0.78
0	0.50	0.42
0.5	0.51	0.02
1.0	0.22	-0.58
1.5	-0.38	-1.20
2.0	-1.00	-1.24
3.0	-1.62	-0.62
4.0	-1.93	-0.31
5.0	-2.18	-0.25
6.0	-2.41	-0.23
7.0	-2.61	-0.20
8.0	-2.80	-0.19
9.0	-2.97	-0.17
10.0	-3.14	-0.17

Graph the data,

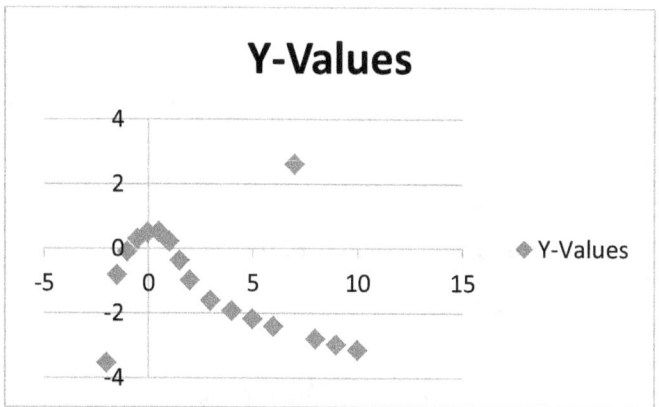

The graph shows the interaction of two functions.

From $-2 \leq T \leq 2$ a parabolic function

From $T \geq 5$ a linear function

Using a TI calculator,

$$Y_{-5 \text{ to } 1.0} = -.5T^2 + .21T + .52, \quad r^2 = .99$$

$$Y_{5 \text{ to } 10} = -.19T - 1.26, \quad\quad r^2 = .99$$

Separately these fit the data but the two functions cannot be combined into one function. This is because the parabolic function dominates the linear function. The reader can add them and view the plot and see the power function overwhelms the linear function

We next go to the linear portion to find the DE.

$\Delta Y/\Delta T = f(Y,T)$
let T = 7
-.2 = Y – 7
Y = 6,81

but Y = 2.61, however, $\sqrt{6.81} = 2.61$

therefore, $dY/dT = Y^2 - T$

If you are using a TI-89 input this DE for the slope field graph and select $Y_1 = 0.5$. You will get the original data set. Now select $T_1 = 1$
You now see Y is some cubic function. Depending on the initial conditions you can have a linear, quadratic, or cubic function. There is no single integral solution to this DE. Select initial condition for Y_1 to be 1.5, -1 and the graph will clearly show this to be true.

The next step is to use a numerical method to calculate the values of Y based upon the given initial conditions.

I will write two numerical methods for the TI calculators: one for the Euler method and one for the Runge-Kutta method.

: Program Euler
: Disp "Input X"
:Input X
: Disp "Input Y"
: Input Y
: Disp "Input H"
: Input H
: LBL 1
: Y + H*(1.0Y(1 - .1Y)→ Y   (Insert here the DE
                              to be evaluated)
: X+H → X
: Disp X,Y
: Pause
: Goto 1
: End

The DE inserted in the line after LBL 1 is for the logistic equation studied previously. For initial conditions insert X = 1, Y = .2 and H = 0.1.

A more accurate, but more complex program, is the Runge-Kutta method listed below.

```
Program:RK
:Disp "Input X"
:Input X
:Disp "Input Y"
:Input Y
:Disp "Input H"
:Input H
:H → N
:LBL 1
:(Y² – X) → D   :Insert here the DE just studied
: H*D → J
 :(X + H/2)) → A:(Y +(J/2)) → B:H*(B²- A)→K
:(X + (H/2)) → A:(Y + (K/2)) →B:H*(B²-A)→L
:(X+H)→A:(Y+L)→B:H*((B²-A)→M
:Y+(1/6*(J+2*K+2*L+M)→Y
:X+H→X
:(N/.5)→A
:N+H→N
:If A≤ 1
:Goto 1
:H→N
:Disp X,Y
:Pause
:Goto 1
:End
```

Note that B = Y and A = X. If your DE was

$$dY/dX = 3X^2 \text{ you enter}$$

$$3X^2 \rightarrow D$$

$$H*3A^2 \rightarrow K$$

$$H*3A^2 \rightarrow L$$

$$H*3A^2 \rightarrow M$$

Another example is,

$$dY/dX = Y - X$$

Enter

$$(Y - X) \rightarrow D$$

$$H*(B - A) \rightarrow K$$

$$H*(B - A) \rightarrow L$$

$$H*(B - A) \rightarrow M$$

When evaluated this last example describes

$$Y - e^X + X + 1$$

Most DE equations cannot be solved directly. These numerical methods allow you to evaluate your DE. But that can also be the problem. Can you always find the DE from your experimental data? The mixed functions just discussed show how thus can be a problem. In the next chapter we will study a more difficult problem, the second order differential equation.

Chapter 11. The Second Order Differential
Equation

Having developed numerical methods to approximate
solutions of a differential equation we can now study
second order DEs.

There are two basic approaches to solving a DE. The
first is to propose a DE based upon some theoretical
ground relating the variables, solve it and check if the
resulting functions provide realistic predictions. The
other is to gather experimental data and from it
develop a DE that yields a function that fits the data.

All appears straight-forward until one attempts to write
the DE either from theory or experimental
data. Fortunately, some experimental data leads to a
straight forward analysis leading to a
solution. However, the solution of most DEs requires
numerical approximation methods but with today's
computers and even graphical calculators this is no
problem.

An illustration of these difficulties and an approach to
finding the DE from the data now follows.

Given the following data,

X	Y
-5.0	.088
-4.5	.133
-4.0	.201
-3.5	.302
-3.0	.448
-2.5	.657
-2.0	.947
-1.5	1.339
-1.0	1.835
-0.5	2.426
0.0	3.000
0.5	3.297
1.0	2.718
1.5	0.000
2.0	-7.389
2.5	-24.365
3.0	-60.257
3.5	-132.462
4.0	-272.991
4.5	-540.103
5.0	-1038.892

Graph the data next,

Figure 11.1

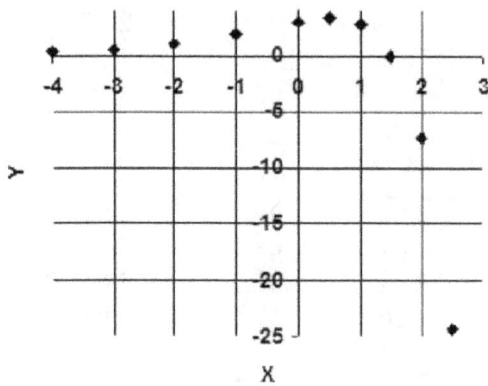

The graph says,

At X = 1, Y = 2.718 = e : possible
exponential function

The second quadrant appears
exponential up to X = 0.5

The first and fourth quadrants appear
influenced by a different function

We, therefore, calculate $\Delta Y/\Delta X$ and ln Y

X	Y	ΔY/ΔX	ln Y
-5.0	.088		-2.43
-4.5	.133	.090	-2.02
-4.0	.201	.136	-1.60
-3.5	.302	.202	-1.20
-3.0	.448	.292	-0.80
-2.5	.657	.418	-0.42
-2.0	.947	.580	-0.05
-1.5	1.339	.784	0.29
-1.0	1.839	1.00	0.61
-0.5	2.426	1.17	0.87

Regressing lnY versus X on a TI calculator gives,

$$\ln Y = .74X + 1.27 \quad r^2 = .996$$

Integrating,

$$Y = e^{.74X} e^c$$

At X= 0, Y = 3 gives $e^c = 3$

Therefore,

$$Y = 3e^{.74X}$$

Keeping in mind the broad divisions of ΔY/ΔX and my usual preference for simplifying, let

$$Y = 3e^X$$

If you compare the values of this function with the data from X = -5 to X = -1 you will find a very poor fit to the data. This is not surprising since we have not yet accounted for the rapid decrease in Y when X is greater than 1.

Now comes the most important part. The number 3 in the equation is due to the initial conditions that resulted in the data we are studying. It is important to generalize
the equation.

$$Y = ke^X$$

Now we find k by regressing X vs. $k=Y/e^X$.

X	Y	$k = Y/e^X$
-4.5	.133	12
-4.0	.201	11
-3.5	.302	10
-3.0	.448	9
-2.5	.657	8
-2.0	.947	7
-1.5	1.339	6
-1.0	1.839	5

A TI calculator regression gives,

$$(1) \quad Y = (3 - 2X)e^X$$

$$(2) \quad dY/dX = (1 - 2X)e^X$$

This can be run on the TI Euler program with the interval spacing, H, set to .01 and a good match to the original data will be found.

Equation (1) gives a function that fits the data. Equation (2) gives a DE that gives good results using numerical methods. But to learn what type of DE you have you have to go to the next step.

First find the second derivative,

$$(3) \quad dY^2/dX_2 = -e^X - 2Xe^X$$

Since the second derivative is neither a constant or zero we go to the next step to see if we have a second order DE.

The general equation for a second order DE is,

$$Y'' + aY' + Y = 0 \text{ or } F$$

We now need to see if equations (1),(2), and (3) can be arranged so they either sum to zero or a constant.

First add the equations Y'' and Y

$$-e^X - 2Xe^X + 3e^X - 2Xe^X = 2e^X - 4Xe^X$$

From this sum subtract 2Y′

$$2e^X - 4Xe^X - 2e^X + 4Xe^X = 0$$

Therefore,

$$Y'' - 2Y' + Y = 0$$

And we have solved a second order homogenous differential equation.

I have reversed the textbook approach where you are given

$$Y'' - 2Y' + Y = 0$$

and the solution follows the pattern

$$r^2 - 2r + 1 = 0$$

$$(r - 1)(r - 1) = 0$$

From this,

$$Y = Ae^X - 2Xe^X$$

with initial conditions $Y(0) = 3$ and $Y'(0) = 1$

so that

$$Y = 3e^X - 2Xe^X$$

What I am demonstrating is the difficulty in determining from the data if a second order differential equation is called for. But it can be solved directly from the original, or experimental data, and the DE calculated after the function is derived.

Chapter12. A Method For Solving
Partial Differential Equations

Except for the chapter describing multiple regression, all of the equations have had the dependent variable, Y, as a function of a single independent variable, X. In many cases, however, the Y variable is a function of two or more independent variables. How does one decide whether a differential equation or a multiple regression is suitable? Two examples should answer the question.

Let's assume you have conducted an experiment on a gas where the pressure has been held constant and the temperature varied. Next, you vary the pressure while you hold the temperature constant. The resultant data collected can be described by a partial differential equation (PDE). When one or more independent variables can be held constant while another independent variable is varied, a PDE may be applicable.

Now you want to investigate some possible factors that may affect the price of stock market shares, e.g., the Standard & Poor's 500 stock index. You want to see if stock prices are directly related to corporations' profits and inversely related to the prime interest rate. But it is not possible to hold corporations' profits fixed at one level while over time you vary the interest rate while you measure the S&P 500 index. This requires a statistical multiple regression study.

To see how to work with a PDE assume your dependent variable is a function of X and Y and you have conducted an experiment that has given you the following data:

Y held constant

X	Y	Z	$\Delta Z/\Delta X$	$\Delta^2 Z/\Delta X^2$
-3	1	-70		
-2	1	-27	43	
-1	1	- 8	19	-24
0	1	-1	7	-12
1	1	6	7	0
2	1	25	19	12
3	1	68	43	24

$$\partial^2 Z/\partial X^2 = (24-12)/(3-2)*X + C = 12X+C$$
$$\partial Z/\partial X = 6X^2$$

$$Z = 2X^3$$

At this time ignore the C terms. They will be
collected into the final term.

Repeat the experiment holding the X term
constant.

X	Y	Z	$\Delta Z/\Delta Y$	$\Delta^2 Z/\Delta Y^2$
1	-2	30		
1	-1	8	-22	
1	0	2	-6	16
1	1	6	4	10
1	2	14	8	4
1	3	20	6	-2
1	4	18	-2	-8
1	5	2	-14	-14
1	6	-34	-36	-20

$$\Delta^2 Z/\Delta Y^2 = ((10 - 4)/(1 - 2))Y = -6Y$$
$$\Delta Z/\Delta Y = -3Y^2$$
$$Z = -Y^3$$

Adding to the previously found Z we have,

$$Z = 2X^3 - Y^3$$

But if we substitute X = 1 and various Y values into this formula we do not get Z' values close to the Z values in the above table.

Therefore, there is some unknown value not accounted for. Let it be identified by the symbol U, so that

$$Z' = 2X^3 - Y^3 + U$$

X	Y	Z	Z'	U = Z – Z'
1	-2	30	10	20
1	-1	8	3	5
1	0	2	2	0
1	1	6	1	5
1	2	14	-6	20
1	3	20	-25	45
1	4	18	-62	80
1	5	2	-123	125
1	6	-34	-214	180

Note that U = a*f(X,Y)

We know that X= 1
What is Y = ?

Take the first row in the above table,

$a(1)Y = 20$

From the second row where Y = -1

$a(-1) = 5$

$5XY = 20$

Since X = 1,

$5Y = 20$

$Y = 4$

But the first row says Y = -2

For this to be true Y must be Y^2.

Therefore, $U = 5XY^2$
And, $Z = 2X^3 + 5XY^2 - Y^3$

To write the results in terms of a partial differential equation recall,

$\partial^2 Z/\partial X^2 = 12X$ and $\partial^2 Z/\partial Y^2 = -6X$

Therefore, $\partial^2 Z/\partial X^2 + 2\partial^2/\partial Y^2 = 0$

is our partial differential equation that when solved gave

$$Z = 2X^3 + 5XY^2 - Y^3$$

The following is another more interesting example.

Given the data.

X	Y	Z	X	Y	Z
1	-3	-26	-3	1	-18
1	-2	-11	-2	1	-2
1	-1	-2	-1	1	2
1	0	1	0	1	0
1	1	-2	1	1	-2
1	2	-11	2	1	2
1	3	-26	3	1	18

The columns of 1's for X and Y indicates a partial differential equation may describe the data.

Plot the data when X is held constant.

Figure 12.1

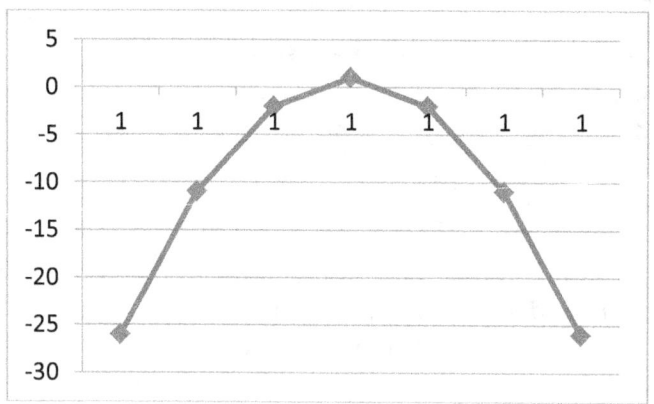

Using the TI calculator and solving for a quadratic gives,

$$Z = -3Y^2 + 1$$

$$dZ/dY = -6Y$$

$$d^2Z/dY^2 = -6$$

Plotting the data set when Y is held constant,

Figure 12.2

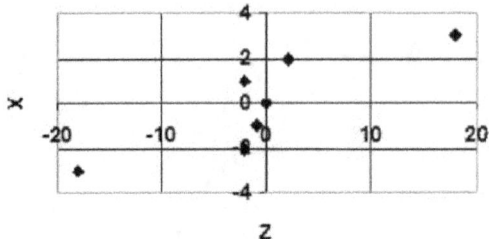

Z

Because of the discontinuity at X = 0, regress
X = -3 to X = -1 and X =1 to X = 3

X = -3 to –1	X = 1 to 3

$Z = -6X^2 -14X -6$ $Z = 6X^2 -14X + 6$

$Z = -2(3X^2 + 7X -3)$ $Z = 2(3X^2 -7X + 3)$

$dZ/dX = 6X + 7$ $dZ/dX = 6X -7$

$d^2Z/dX^2 = 6$ $d^2Z/dX^2 = 6$

Since the second derivatives are the same for both data sets we can write the partial differential equation as

$$\partial^2 Z/\partial X^2 + \partial^2 Z/\partial Y^2 = 6 - 6 = 0$$

This is the famous Laplace Equation in 2 dimensions.

Now we need to solve this Laplace Equation to find

$$Z = f(X.Y)$$

Proceeding as in the previous example, from the table where Y is held constant:

X	Y	Z	$\Delta Z/\Delta X$	$\Delta^2 Z/\Delta X^2$
-3	1	-18		
-2	1	-2	16	
-1	1	2	4	-12
0	1	0	-2	-6
1	1	-2	-2	0
2	1	2	4	6
3	1	18	16	12

Graphing,

Figure 12.3

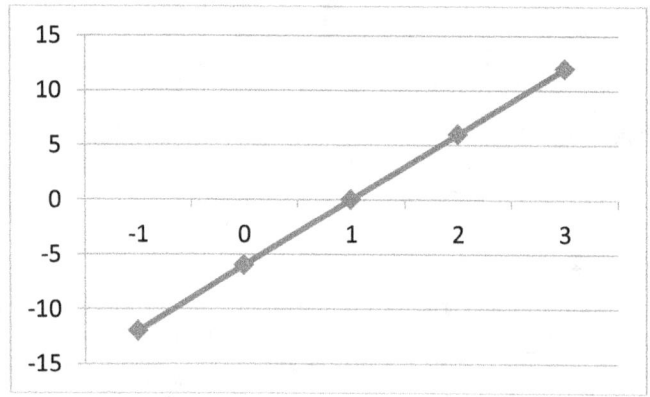

Resulting in the equations,

$$\Delta^2 Z/\Delta X^2 = 6X$$

$$\Delta Z/\Delta X = 3X^2$$

$$Z = X^3, \text{ ignore constant terms}$$

Now let X be held constant.

X	Y	Z	$\Delta Z/\Delta Y$	$\Delta^2 Z/\Delta Y^2$
1	-3	-26		
1	-2	-11	15	
1	-1	- 2	9	-6
1	0	1	3	-6
1	1	-2	-3	-6
1	2	-11	-9	-6
1	3	-26	-15	-6

Graphing,

Figure 12.4

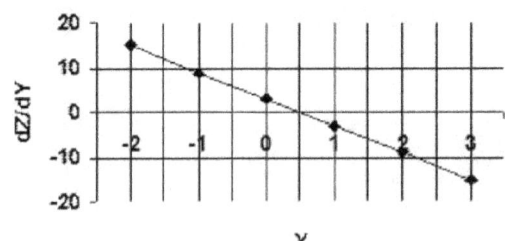

Solving,

$$\Delta Z/\Delta Y = -6Y + 3$$
$$Z = -3Y^2$$

Combining,

$$Z' = X^3 - 3Y^2$$

Check the results of Z' versus Z.

X	Y	Z	Z'	Z – Z'
1	-3	-26	-26	0
1	-2	-11	-11	0
1	-1	-2	-2	0
1	0	1	1	0
1	1	-2	-2	0
1	2	-11	-11	0
1	3	-26	-26	0
-3	1	-18	-30	12
-2	1	-2	-11	9
-1	1	2	-4	6
0	1	0	-3	3
1	1	-2	-2	0
2	1	2	5	-3
3	1	18	24	-6

This indicates the $-3Y^2$ needs modified. Since the interaction with Y is not independent of Y let U equal the interacting factor and solve for U.

$$Z = X^3 - 3Y^2(U)$$

$$Z - X^3 = -3Y^2(U)$$

$$U = (Z - X^3)/-3Y^2$$

Solving for U,

X	Y	Z	U
-3	1	-18	-3
-2	1	-2	-2
-1	1	2	-1
0	1	0	0
1	1	-2	1
2	1	2	2
3	1	18	3

Note that,
$$U = X$$

We can now write the solution to this particular Laplace equation as,

$$X = X^3 - 3XY^2$$

It has been shown that by using experimental data and the methods developed here for solving differential equations we can derive and solve partial differential equations.

The question to ask is, "When is the U term additive or multiplicative?" To our general method of graphing, find the derivatives, and integrating, must be added:

If $(Z - Z')$ is not equal to zero for both for   both independent variables the U term is additive.

If $(Z - Z')$ equals zero for at least one of the independent variables the U term is to be multiplied to the variable held constant when $(Z - Z')$ is not zero.

Some interesting questions to solve are:

Do all Laplace equations form a group where one or more independent variables have $(Z - Z')$ equal zero?

What if the sum of the second order partial derivatives does not equal zero. If so, is this an extension of the above mentioned group?

Chapter 13. A Final Caution

I have stressed repeatedly the danger of extrapolation. Remember a polynomial has a number of roots equal to the degree of the equation. If you extrapolate beyond your data you may find your derived equation suddenly diving down to the X axis. Unless the theory underlying your derived equation can account for the degree of your derived equation, polynomials are best used for interpolation values between the data points.

You will often come upon the word "trivial". But what does this word actually mean when applied to a mathematical result? And how can you judge whether your result is "trivial" or not? I will give you a very clear example to demonstrate "trivial" or not.

There have been several popular books written recently about the gamma function[1]. The discovery of the gamma function came from attempting to find the fractional values of a factorial, e.g., 5.5! I will now derive a trivial solution of 5.5!

First set up a data table and the draw a graph.

X	Y = X!
1	1
2	2
3	6
4	24
5	120
6	720

Figure 13.1

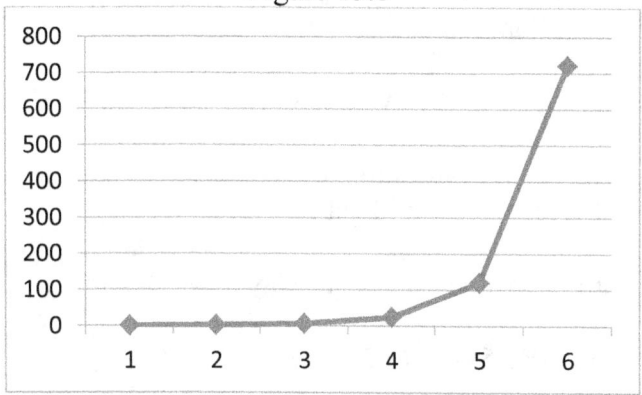

We can now use the methods we have learned to find the value of the factorial 5.5.

Let's first try a quadratic.

X	4	5	6
X!	24	120	720

Set up three simultaneous equations,

$$a(4)^2 + b(4) + c = 24$$
$$a(5)^2 + b(5) + c = 120$$
$$a(6)^2 + b(6) + c = 720$$

Using the matrix function on a calculator gives the following solution,

$$X! = 252X^2 - 2172X + 4680$$

As expected, the X! values for X's 4,5, and 6 are right on. We solve for x = 5.5 and get X! = 357

Are we sure this value is correct? Maybe we should try

$$(1.5)(2.5)(3.5)(4.5)(5.5) = 324.84$$

Or should it be half this

$$(0.5)(1.5)(2.5)(3.5)(4.5)(5.5) = 162.42$$

Something is very wrong!

Let's try a fifth degree polynomial.

$$a(1)^5 + b(1)^4 + c(1)^3 + d(1)^2 +$$
$$e(1) = 1$$
$$a(2)^5 + b(2)^4 + c(2)^3 + d(2)^2 + e(2) = 2$$
$$a(3)^5 + b(3)^4 + c(3)^3 + d(3)^2 + e(3) = 6$$
$$a(4)^5 + b(4)^4 + c(4)^3 + d(4)^2 + e(4) = 24$$
$$a(5)^5 + b(5)^4 + c(5)^3 + d(5)^2 + e(5) = 120$$
$$a(6)^5 + b(6)^4 + c(6)^3 + d(6)^2 + e(6) = 720$$

Again using the TI calculator matrix function to solve,

$$X! = 2.57X^5 - 36.42X^4 + 198.63X^3 - 511.53X^2$$
$$+611.8X - 264$$

This is certainly a formidable equation.
Let X = 5.5 and

$$(5.5)! = 282.33$$

Which value is correct? Is any value correct?

Have we derived these values from any underlying theory? No. We just mechanically derived formulas without any prior analysis.

It was Euler's work on deriving analytically a function to calculate factorial values that led to the creation of the gamma function. The two books previously referenced give an excellent account of how the gamma function was derived.

My purpose is to show the results and how it was used to find 5.5!

Euler's formula for calculating factorials is,

$$n! = \int X^n (e^{-X})$$

The integration in this formula is taken from X=0 to infinity.

Using the TI calculator enter into the Y editor, solving for 4!

$$Y = X^4 (e^{-X})$$

In Graph mode calculate the integral from X = 0 to X= 20 and get

$$4! = 23.999593 \text{ versus actual} = 24.$$

Finally,

$$5.5! = 287.84548$$

The derivation of the gamma function is long and difficult but it demonstrates the key point about what is "trivial". Euler didn't want an interpolation formula. Instead, he derived his equation by fundamental analysis of the problem.
So does this explain what is mathematically trivial?
Euler's work was brilliantly derived

based upon the insight of genius and analysis. But the mechanically derived formulas above are trivial. They are as meaningless as a researcher adding more terms to a regression equation just to achieve a higher r^2.

We have now come full circle. In the first chapter it was stressed that many equations can be derived to fit a set of data. The properly derived equation must be based upon a theory or underling understanding of what variables are reasonably expected to interact. The derived equation is a mathematical expression of this interaction. Without this understanding there will be a good chance your equation has little merit.

But I will close, however, on a positive note. Many powerful methods of treating data have been learned from elementary quadratic to differential equations. These are powerful tools to analyze experimental results.

1. John Derbyshire, "Prime Obsession", Joseph Henry Press,
2003

2. Julian Havil, "Gamma: Exploring Euler's Constant", Princeton University Press,2003
3. Philip J. Davis,"Leaonhard Euler's Integral: A Historical Profile Of The Gamma Function", J. C. Abbott, editor. Mathematical Association Of America,1978.

---

[1] "Differential Equation Models",Martin Braun editor. P.40.
Springer=Verlag,1983